THE EMBRACING WOODS:

A Book for Fathers To Give Their Sons

by
Richard Lee Fulgham

Text and Cover Copyright 2003 by the author

LION PUBLISHING, INC.

www.lulu.com/LION

All Rights Reserved

Printed in the United States of America

1 2 3 4 5 6 7 8 9 0

"If I remember the forest it is because from its hidden reaches man arose. The green world is his sacred center. In moments of sanity he must still seek refuge there."

— Loren Eiseley

"Loren Eiseley would have loved this book. It's a nature book for nature lovers. The Embracing Forest concerns a young naturalist growing up in a small 1950's town snuggled up to Georgia's Pine Mountain, part of the Great Pine Forest. The book remembers the boys, beasts and bullies -- the escapes into the woods to find peace and wisdom. It's about animal behavior observed and the boy's realization that animal behavior and human behavior are intimately related -- indeed the same. It's also beautifully written (see sample chapter.) God bless the boys and beasts of Fulgham's pine forest."

-- Roger de Rageot,
Dr. Sci. (Sorbonne) Professor
Emeritus, former Curator,
Norfolk Museum of Natural
History.

Table of Contents

The Great Pine Forest

Quicksand

The Alligator and I

The Day of the Black Boar

The Goat Man

The Bat with a Human Face

The Goldenrod Spider

Fear in the Desert Night

The Horn Hunter

Fox and Hounds

The Death of the Viking Princess

The Courage of Field Mice

Wisdom in a Serpent's Eye

Chapter One
<u>The Great Pine Forest</u>

Georgia's Pine Mountain is the southernmost ridge of the mighty Appalachian Range, stretching from the Blue Ridge all the way down to the Chattahoochee River -- from above Atlanta to below Columbus. Comprised of quartzite, it's heavily forested with pines and bespattered with rock outcrops. It's interlaced with clear streams and sparkles with tumbling waterfalls.

The mountain is not very high, only an hour's climb for a kid, maybe two for a short-winded adult. But it's worth the effort because from its crest you can see the Great Pine Forest extending horizon to horizon in all directions, carpeting the land with millions of loblolly pines and kudzu vines. You can see the ocean of evergreens covering west central Georgia – the immense woods in which a kid can explore forever into the hidden and unexpected worlds of Nature.

Into this sea I once escaped, when I was maybe nine or ten, the small town life I found so boring and dull. The time was the early 1960's and the town was called Manchester – a cluster of about 3000 souls at the foot of the mountain. I wouldn't be so bored today – having suffered so many strenuous times in my adulthood that I desperately crave the peace and quiet I once despised.

How sad that it's now too late to find that state of grace again. There's no returning. Those days have been absorbed by time and the small town has grown into a city in which I am a stranger. I return only to roam the streets like a ghost, invisible to the young and unrecognized by the old. Only a few souls who knew me still live there. But they are no longer the kids I knew when we were wild and free.

Thus laments a native son . . . .

I have tried to return but nowhere can I find the Main Street I remember, the avenues I walked, the alleyways I explored, the gang I knew, the trails I tramped, the dogs I loved. The town I knew is gone forever. But I do have one consolation -- the mountain and forest remain as they were, vast and wild. The mountain still welcomes any lonesome souls who crave the solitude of her crest. The great pine forest still whispers strange and profound lessons to those who wander along her trails. The wisdom of the woods is still to be found.

I resist being personal but must dip a little into that time forty, fifty years ago when other kids and I could roam without limits or licenses or fears. It was a time when dogs ran free and so did we. Only the schools we attended – Manchester Grammar and Manchester High – penned us in. But even those we could escape every day at three, rushing home to change our school attire for the worn out clothes our mothers refused to throw away.

Of course, we never hung around our

homes – for us it was the street or the woods. The outdoors was our reality. We thought of ourselves as prisoners when forced to stay at home or trapped at school. We roamed as a pack in our preteen years but when adolescence came I separated myself and turned more and more toward the woods, preferring the mountain over the baseball diamond, the trees over my pals, the animals over the adults. By the time I was thirteen, I was a bonafide, dyed-in-the-wool loner.

At that time, in the mid 1950's, Nature and humanity seemed two different forms of life to me. I can't speak for my old buddies but decades would pass before I would realize that the town and its people were as much a part of Nature as the forest and its creatures. How could I have understood then that the incomprehensible forces which ruled the woods also ruled the township – that the nature of the wild was also the nature of man?

Though I had plenty of pals, I was always alone in my heart – so most days at four o'clock I would leave the house by myself to follow a trail I'd made over the mountain and through the woods. It led to my secret refuge – a patch of old pines carpeted with soft, dry pine needles and drained by a tiny stream I called Copperhead Creek. It was clear as only a Georgia creek can be.

It was about five feet across but only a few inches deep as it flowed through the refuge. But I could take off my shoes and socks and wade a mile or so down till I found pools where the water had collected in ravines. I named one of these pools – one especially full of snakes – Moccasin Pond.

The water had pooled in a deep trench cut through red clay. It was about ten feet in length, four or five feet across, about two feet deep and was bordered on both shores by perpendicular clay shelves rising five or six feet from the surface of the water. Someone years before had tried to bridge the ravine with an old wooden door, which had eventually fallen into the little gorge lengthwise to make a platform over one end of the pool.

I was able to stretch out on the old door and peer directly into the pool. The ravine was wide enough so that the sun almost always flooded my private pond with a penetrating white light which made the water and everything in it shine with an awesome clarity. Vines crisscrossed the tiny realm over my head and dozens of queen watersnakes sprawled across them, basking in the sun.

It was a microcosmic, magic world. But it was not mine alone, for I shared it with queen snakes, green snakes, garter snakes, ribbon snakes, brown snakes, a massive tub-sized snapping turtle, a smaller hand-sized mud turtle, three painted terrapins, a box turtle, hundreds of anole lizards, a half dozen spiny swifts, one five-lined skink, six brown skinks, one grapefruit-sized bullfrog, five green frogs, dozens of smaller leopard frogs, a slew of peeping treefrogs, countless salamanders and their aquatic young, and countless mudpuppies, tadpoles, minnows, sunfish, crayfish, dragonflies, damselflies, and other monstrous-looking insects and their larvae.

All lived together in the crystal creek. Oh yes – there was also a young water moccasin who always basked alone on a particular branch, far away from the other watersnakes – and a coral snake, whom I saw only once.

Upstream, I knew, lived a giant six-foot brown watersnake and somewhere in the woods lived a cannibalistic kingsnake. I had seen them both but they rarely visited Moccasin Pond when I was around. There was also a sly gray fox whom I'd seen only once, surprising him as he devoured a squirrel he'd ambushed. Over the pool and all around lived thousands of birds, including a barred owl who slept in a hollowed out cavity in a huge dead tree and a kingfisher who buzzed my pond every day. So this was my world and its exquisite sea.

And therein lies a tale . . . .

One Saturday morning, just after the sun had risen as white and round as the eyelid of a dying bird, I lay on my bridge and was watching the mudpuppies stalk earthworms I had dropped into the water. The turtles and crayfish were hungry too, and a little battle erupted beneath the surface as the predators fought each other for the drowning worms.

The commotion was so great, and I so thrilled, that I didn't notice a fat queen watersnake crawling out of the water and onto the bank. The bank was covered with muddy grasses and it was the sound of slithering that awoke me from my transfixed state. Looking at her, I saw that she was about two feet long and apparently searching for something.

Suddenly she seemed to find the right place, next to a large piece of rotted log, and began to shiver as if freezing. Then, to my astonishment, she began giving birth to living young. I knew that some snakes lay eggs and some give birth to living young. But reading is one kind of excitement and actually witnessing such an event quite another. One by one the baby watersnakes were born, each wrapped in a transparent membrane, through which they easily tore.

She gave birth to seven snakes, each about four or five inches long and as big around as a pencil. As they broke through the membrane, each would instinctively head for the water. The mother paid no attention to them. She didn't care about their fates.

As I watched, hypnotized, the hideous snapping turtle appeared and snatched the first baby snake to slide into the water. As the gruesome snapper grabbed it, the baby seemed to give a silent scream, opening its mouth wide and writhing back and forth as blood tinted the water red. It was over quickly and I watched as the monster turtle dragged its victim into its lair dug into the bank underneath the water.

The second baby watersnake slid into the water and immediately disappeared, safely hidden underneath the vegetation along the shore. I was happy for it.

But Nature had a lesson to teach me. I was

caught by surprise as the big kingsnake appeared from nowhere and began swallowing the baby watersnakes one after another until none at all were left. The aptly named king, gleaming black with a golden chain shining on its back, didn't bother to kill them first. The king merely grabbed a tiny head and swallowed the tiny body in a matter of seconds. And when he was finished with the young, he began to flick its tongue excitedly over the mother, who finally came to her senses and tried to escape.

It was in vain. The kingsnake had her before she could hit the water, grabbing her just behind the head and coiling around her in a ball of rippling muscles and glittering scales. But she didn't accept her fate lightly, thrashing and biting for a full five minutes before she at last fell limp and lifeless, to be swallowed whole by the king of Copperhead's Retreat.

I was paralyzed with excitement and awe. How swiftly and cruelly death came in the wild! Life was extinguished almost as quickly as it was born. Out of seven young and the mother, only one had escaped to perhaps grow to maturity and give birth again to keep the queen watersnake population alive.

It was then I had my first inkling of Nature's first rule: produce many so that a few may survive. At that time, child that I was, I didn't see that this same rule applied to mankind and to mankind's ways. I would not only understand when I reached adulthood, I would experience the same unbending rule in my own life and ambitions.

It was early spring and the pool was filled

with clear, gelatinous blobs of toad, frog and salamander eggs. There were literally thousands of tiny black eggs in the blobs, deposited by the adult amphibians after mating in February as they emerged from hibernation.

I watched those eggs day by day as they changed from black dots to true embryos. I took some home at intervals to study under my microscope. Even then, at that young age, I knew from my books that all embryos look startlingly alike at the beginning. In fact, even human embryos look like the frog and salamander embryos at the earliest stages, with tails, gills, and red, beating hearts seen through transparent skin.

As the weeks passed, the eggs began to hatch and fill the pool with thousands of tiny tadpoles, most of whom were eaten by larger creatures as soon as they tried to swim. I admittedly felt bad about this carnage occurring in my little universe. As I watched, the tiny infant creatures were mercilessly devoured by the turtles, mudpuppies, crayfish, and especially the hideous dragonfly larvae with their long, fanged lower lips. I came every day -- and every day there were fewer and fewer tadpoles. By summer, there were only a hundred or so left out of the hundred thousand or so which had been born.

As the surviving tadpoles increased in size and began to develop limbs, they became prey for the snappers and the brown watersnake, which would come only when I was gone. Only a few dozen tadpoles and salamander larvae remained by the time summer arrived. But even those were endangered. An unexpected threat awaited them on land.

Many, many times as I was watching, a tadpole which had survived long enough to transform into a true frog or toad would boldly hop on land and see for the first time the terrestrial jungle of grasses and weeds it would inhabit. But as dusk came, the mighty bullfrog would take his position on the shore and, between deep-throated songs, would snap them up as if they were insects. To make things even worse, a gartersnake appeared and also feasted on the young amphibians. As each died, it gave a tiny cry of fright and horror as it disappeared down the carnivore's throat. Maybe a dozen lucky infant frogs and toads escaped into the night.

So, out of at least 100,000 new lives, about 12 survived. It was embedded in my mind: Nature creates thousands or millions so a few might survive and reproduce. Such is the struggle to survive. Next year there would be maybe five or six new frogs living on the shores of Copperhead's Retreat. They would sing into the night and feast and copulate and live until found by the snakes, bullfrogs, water birds or other predators hunting their flesh. Just to live was to live dangerously.

I have never forgotten the carnage I saw at that tiny pool of water in the middle of the pine woods. So I was not surprised when war and accidents and disease took a toll of my friends as I matured through the years,. I had grieved for snakes and frogs as a child; as an adult I grieved for my lost friends. But I understood and did not condemn life or accuse God of cruelty. Nature feeds upon itself and in this manner grows. It could be no other way.

Though I went through an adolescent stage when I blamed mankind for all the pain and death in this too real world, I eventually connected my lessons at Copperhead's Retreat to the process of existence. Nature is one immense organism growing in the Universe and those creatures we call individuals are cells within its body. As in all bodies, cells mature, reproduce, then must die to make way for the young.

The universe is a cold and hostile place. Nature must make up for this by over-producing in order to appease that higher consciousness ordering the grand scheme of being. Nature is lush, fecund and beautiful -- but she can only be this way by constantly creating in such abundance that death cannot devour all of her creations. Whether there is a God separate of Nature, or whether God is Nature, the same basic rule applies and can't be changed.

Now, having reached that point in my life where I see more behind me than I can see in front, I understand and accept the violence that seems to engulf this living earth. The fear, grief and pain we must experience is not cruel. It is only natural and in the long run, renews and refreshes Nature as she feeds off of the dead.

And I understand, too, that there must be a single underlying spirit animating all living things, a shared spirit which never dies with the material flesh but continues forever as Nature proceeds into the unknown tomorrow of the future. Perhaps in the end, Nature will indeed be one vast living, conscious being which will survive and reproduce until the universe is filled with life.

There was much more for me to learn from Nature, and we shall discuss these hard-earned lessons in coming chapters. In the words of the existential philosopher Jean Paul Sartre, "Nature speaks and experience translates. Men have only to keep their mouths shut." I have lived by this advice when in the wild, and it is only now, among my fellow humans, that I try to translate Nature into an understandable human language.

Despite Sartre's well meaning advice, there comes a time when we must try to communicate the will of Nature to each other -- and in that process come to understand that life and death are intimately married and neither is to be feared, only accepted. With this understanding and acceptance, perhaps we eventually will leap through life like laughing lions, strong, bold, wise and free.

Chapter Two
<u>Quicksand</u>

In their efforts to make Manchester, Georgia, a cleaner and safer town, the administration of 1961 decided to drain the city lake and install an overflow device. The idea was to prevent flooding, though in fact the pond had never overflowed in the history of the community.

Manchester's Mayor Boss Roundtree, Sr., with the blessing of Big Jack Middleton – Sheriff of Meriwether County – thereby authorized a bulldozer to slice a huge ditch in the lower bank, allowing all the water to gush out, complete with snakes, terrapins, snappers, soft-shelled turtles, alligator gars, big mouth bass, small mouth bass, bluegill, bream, perch, suckers, catfish, frogs, salamanders, crayfish, and several million aquatic insects.

The resulting flood caused the little creek, which the pond had formerly used as an escape valve, to swell into a torrent of mud, water, and creatures gasping for life. It was the worst catastrophe I had ever seen and thousands of animals died that day.

The emptied lake wasn't refilled for another year, so for 12 months we who had loved the pond saw only a dried crater layered with the decaying bodies of fish and turtles. The putrid odor was enough to gag a maggot. The turkey buzzards came every day at first but eventually not even they could stand the rotting cadavers.

But Nature, like God, works in mysterious ways. An overflow pipe was installed underneath the earthen bank, which was supposed to drain off excess water and let it escape through the pipe, under the earth and into the little creek. In other words, it was precisely as it was before draining, only the conduit carried the water underground while the former arrangement had allowed it to escape over the top.

But they discovered an error had been made by the engineers when they finally refilled the pond. The lower end of the drainpipe was quickly covered by layer after layer of sand until it lay three or four feet beneath the creek bed. As a result, the water trickled downward out of the pipe and kept the sand above in a continually fluid state. From a human standpoint, the creek bed looked firm. But if you waded into the creek, you immediately sank up to your knees in liquid sand pulling you further and further down.

In other words, the town officials had in their wisdom created for their constituents a pool of quicksand. Dogs disappeared. Person after person sank in, almost always having to give up their shoes to escape. The reason so many people waded into the creek was to capture minnows and mudpuppies for fishing in the refilled pond. Until Marvin Bozeman, no one knew how deep a person might sink if he or she couldn't escape immediately.

The trap was particularly deceptive and

dangerous, as the quicksand dragged you further and further down the more you struggled. The only way out was to stop moving, get rid of your shoes and lay flat on the creek bed while gently loosening the grip of the sand on your legs. Then you could reach the shore and pull yourself out by grasping water weeds with your hands. All of us kids knew this except Marvin Bozeman.

Like my friends, I discovered the hard way about the quicksand booby trap. I lost a fine pair of Wellington boots and almost lost my dog, whose imaginative name was Rex, and who sank up to his neck after jumping into the creek. Rex was not happy about his plight and howled with indignation as I pulled him out by his tail. He was angry for weeks and never again entered any body of water bigger than the tub in which I washed him. He was even suspicious of that.

Unfortunately, Marvin wasn't suspicious at all. He was like Wimpy in the Popeye cartoons -- pear-shaped and weighing almost 200 pounds. His ass looked like an onion and he walked like a girl. His hobby was the study of stamps and he spent most of his free time at the library with his giant collection arranged by date and country in an absolutely massive stamp-book. He was also extremely conscientious at school, making all A's and joining all the intellectual clubs. The teachers fawned over him and had him sit in front of their classes. He gave eloquent speeches in assembly every Friday. He wore a vest. We didn't like him at all.

Neither did it make him likeable when he finked out on someone who dared argue with him. He always said the same thing when he felt he was losing a debate. "You," he would say with disdain, "are an irretrievable faux paux of another sanctimonious sect, bordering perhaps on the overt periphrasis of a psychic trauma." Then he would stare at his befuddled opponent a few moments and turn his back in disgust.

So it was a source of great annoyance to the rest of us when Marvin joined our Boy Scout troop and quickly beat us all at gaining rank and earning merit badges. He made Wolf before I could make Tenderfoot. He was also in terrific physical shape despite his weight. When we made our twenty-mile hike that year, he was beaming and bouncing at the end of the journey, while the rest of us were panting like dogs and flopping on the ground.

He was infuriatingly superior in every way except in being a cool cat. We were cool, this we knew – we patterned ourselves on Maynard Krebbs, the finger-snapping beatnik on television's "Dobie Gillis". Some of us were even as cool as Kookie on "Sunset Strip". I alas was only semi-cool, basing my coolness on Paladin of "Have Gun Will Travel". He was cool, you see, but he was also a loner – that was why I identified with him. I wore tight black shirts and carried a toy derringer.

Alas, because I was Paladin I never considered myself a real member of the cub scout pack and had joined mostly for the free camping trips. So, when it came time to discuss our campout at the city lake, the others – to my amazement – took me into their confidence and together we cooked up a plot to bring Marvin down. They weren't that happy with my presence there -- but at least I was semi-cool. Marvin wasn't cool at all. He was a square. Yet the scoutmaster had let him in our gang of preteen hipsters. He was in and uncool. He just didn't belong in our neat little pack. No one could have been more despised than this superior in our midst.

Not counting Marvin, there were eight of us: Willy, Roundtree the Mayor's son, Cody, Pootaroot, Jimbo, Jackson, Jack Keefer and myself. The scoutmaster's name was Mr. Rabin, but naturally we couldn't let him in on the plot. He was respected by all of us, not because of his age or wisdom, but because he had lost his left thumb in the Korean War. Sometimes he would pat us with that monkey's paw and it was like being touched by an alien. Whenever I shook hands with him, I immediately made three wishes.

Each of those boys had a character all their own, but here we must deal with the bully called Pootaroot. The second son of a well-to-do family, he was tough, a good fighter, a good athlete, tall, classically handsome, and witty. He had actually kissed a girl! Oh, he was cool, all right . . . but he had a bad habit of provoking fights with we smaller and less handsome boys, usually in front of a crowd. He was a master of humiliation if you refused to fight -- if you chickened out. More than once I had tasted the gall of his sarcasm and the threat of his fists. I was a chicken par excellence.

Much of his behavior can be explained by the animal instinct prodding him to vie for alpha male status, which was most easily earned back then by using brute force and intimidation. But there was more to Pootaroot than a genetic predisposition towards humiliating people like me. I was to learn this during our fateful camping trip at the refilled city pond – the day we planned to lure Marvin into the quicksand.

The Georgia Fish & Game Department had restocked the pond with fish by the time we were planning our camp and I had personally thrown in an unwnated menagerie of water snakes, terrapins, snapping turtles, frogs, tadpoles, insect larvae, and catfish. I had also personally built a campsite in a clearing next to the shore, only a few steps away from the treacherous creek. There was a large depression surrounded by big rocks for a fire, alongside of which lay a large log for a bench.

We set up our camp early Saturday morning – it was in July – and spent the rest of the day fishing. Except Marvin, who preferred to net tiny creatures from the water and examine them through a microscope he had brought for that purpose. Once he shouted to us, "Planaria! I found planaria!"

We ran over to peer through his microscope, only to see little brown flatworms with two dark spots for eyes. We weren't impressed, but Marvin was jubilant and excited. "You can teach a planaria little tricks. Then you can grind him up and feed him to a bunch of other planaria and they'll know the little tricks too! Can you imagine? Memory passed on by digestion! If we could do that, we could grind up Einstein's body, eat him like hamburger, and we'd all be as smart as Einstein! Maybe those cannibals who eat the brains of their enemies knew what they were doing after all?"

We looked at him with new eyes. Then we looked at each other. This guy was a mad scientist in the flesh. We headed back toward our rods and reels by the waterside. We were all perplexed and mystified by what he had said. I decided wrongly that he'd been pulling our legs. But I noticed an expression on Pootaroot's face I had never seen before. It was the expression of a young man who has just learned the fascination of science.

That night there was another curious incident. Willy had found an enormous black beetle the size of a tangerine. He was going to throw it into the fire to watch it explode as the liquids in its body turned into steam. Before he could do it, however, Marvin was between him and the fire.

"Let him go," Marvin said.

"It's just a bug," Willy answered defiantly. "What good is it?"

"What good are you?" Marvin replied, his face darkening with anger. More words were exchanged and the language grew foul. Willy was not a small boy and a fight between him and Marvin would be one hell of a show. Willy threw the bug toward the fire but it managed to take to its wings and buzz away before being consumed. Willy was furious and shoved Marvin, who shoved him right back. The scoutmaster stood up to prevent the coming brawl but before he could speak, Pootaroot was on his feet and between the two combatants.

    Speaking to Willy in a chillingly calm voice, Pootaroot told him to sit down and shut up. He did so immediately, as I would have. Willy was a chicken, too. Surprisingly, however, Marvin gave Pootaroot a tremendous shove, knocking him face down on the grass. Instantly on his feet, Pootaroot gave Marvin a haymaker on the right ear. Then the two were on the ground, with the scoutmaster trying to pull them apart. Us cool guys just watched.

    It was a short but violent fight with no winner. Eventually both boys calmed down and sat back down by the fire, grumbling angrily to themselves. Both promised each other the fight had only begun and with tomorrow would come terrible pain and defeat.

    Without a word, all of us except one decided this night was not the right one for playing a trick on Marvin. We had no idea his temper was so volatile or that he was so quick to react with violence. I was amused by all that had happened but mostly I was happy the giant beetle had escaped unharmed.

    Around midnight under a half moon, Pootaroot told us we were all going catfishing. It was not a suggestion -- it was an order. He was asserting his authority by making sure we had lost none of our fear of him. I didn't care. It was no skin off my ass. He could be the leader if he wanted to. Just so long as he told us to do what we were planning to do anyway. Catfishing was okay by me.

But when Pootaroot announced that we were going to look for crayfish in the creek, we realized the trick on Marvin was still going to happen. Reluctantly, all of us except the scoutmaster headed for the waiting quicksand and inevitable showdown between Pootaroot and Marvin.

But things went wrong. At the shore of the creek, we all pulled off our shoes and rolled up our trousers as if we were going to wade for crayfish. But Marvin just stood by the water's edge, peering at the bottom.

"We can't go in there," he said. "That's quicksand. Notice there's no footprints. No raccoons have been searching for crawdads here. There are no insect trails. No snails. Not even any mussel shells. Nothing. That's a dangerous spot, men."

The jig was up so we began to put our socks and shoes back on. I was glad it hadn't worked. Pootaroot might take it out on us if he got angry enough. But Pootaroot didn't look mad. He looked relieved -- and I understood he had gone on with the plan only to reaffirm his dominant role in the pack.

What no one counted on was that Willy was still angry. We had dismissed and forgotten him. Without warning, he came up behind Marvin and gave him a vicious shove, sending him headfirst into the creek. His head and arms were underneath the sand and sinking deeper before we could react. We were so horrified and shocked we could only stare. Willy laughed once but stopped when he saw the expression on our faces.

Pootaroot snatched off his shoes and jumped into the creek, sinking himself up to his knees. He tried to pull Marvin out but only managed to pull himself down to his waist. Three of us grabbed Pootaroot by the shoulders – which we could barely reach – and managed to free him. But Marvin kept sinking headfirst. We knew he'd be lost forever if that big ass dipped beneath the sand.

Instead of climbing ashore, Pootaroot stretched out flat and instructed us to pull on his legs while he grasped Marvin around the waist. All of us pulled, including Willy and the scoutmaster, who had heard us screaming. Marvin began to rise and eventually popped up like a cork out of a wine bottle. He was coughing up sand and water but very much alive and kicking.

Willy collapsed and cried, mumbling he didn't mean to harm anyone. In a few minutes, however, Marvin and the rest of us were laughing. Willy was forgiven when Marvin sat by him and whispered things we could not hear. Afterwards, Willy came around and shook hands and smiled and joined us again, blushing and ashamed.

But it was Pootaroot who remained silent and somber, perhaps wondering what made him save someone by risking his own life. I couldn't tell what he was thinking, but I could feel what he was feeling. Pride. Self-respect. Wonder. And thankfulness. By saving Marvin, Pootaroot had saved himself. Till that moment when he grasped Marvin's waist, he had lived by a code of the jungle. Suddenly he was aware that there is a part of humanity which is purely human: compassion and self-sacrifice.

Pootaroot received a medal from the Mayor and a big write-up in the town's newspaper. But it only embarrassed him. Something in him had risen above animal instinct. I don't know whatever happened to Pootaroot but I do know that he never again bullied any of us again. And in that way, Pootaroot became a true leader and we felt for him a loyalty and trust untainted by fear. I heard later that he had become an army officer and distinguished himself in Vietnam by discovering an unknown species of butterfly.

Chapter Three
<u>*The Alligator and I*</u>

It was with some trepidation that I agreed to camp out in Georgia's great Okeefenokee Swamp with my Boy Scout Troop. Not because I was afraid of the swamp, but because I was distrustful of the other boys. It was May of 1962.

They were all going. Pootaroot, Roundtree, Marvin, Willy, Jack and four others, plus a new recruit called Gooch. We called him Gooch because his real name was Carol Buchanan and he went berserk when anyone called him Carol, flying at the offender with a rain of fists and atrocious curses. Even the teachers and preachers called him Gooch.

He was a head taller than any of us and looked rather like a stork, being very thin with skinny long legs, pencil neck, and narrow face. His dad was the town's pharmacist, which convinced us all that Gooch was a rich kid. Most of our parents worked at the textile mill, so Gooch represented a financial class far above us. While we wore Sears & Roebuck ® clothes to school, he wore tailored white going-to-church shirts with cufflinks and expensive slacks with buckles in the back.

He treated us with contempt, so naturally we admired him and treated him with great respect. He was the personification of sophistication and class so far as we were concerned, despite his nickname. But there was a line between Gooch and us which we felt should not be crossed. And when Gooch said he was going into the Okeefenokee Swamp with us, he crossed that line. We didn't want any pampered rich kids with us -- he might sneer at our great adventure.

But we all signed up for the two-week campout anyway. It was the chance of a lifetime, as the swamp is a watery labyrinth of phenomenal wonders populated with fantastic prehistoric creatures. If I didn't go with the gang, I'd most likely never go at all. We'd all been told over and over again how the law strictly forbids individual exploration within the swamp. You can only camp in groups at certain places, under the supervision of a park ranger.

This is not without good reason. The swamp covers over one thousand square miles, a dark and ominous expanse of primeval acid-filled black water, ancient cypress trees with odd roots that grow up out of the water like spikes, deceptive islands of quivering "land" which pull you down into their gelatinous muck, thousands of water moccasins and giant alligators, millions of foreboding canals which lead nowhere, and billions of blood-sucking mosquitoes, horseflies, yellow flies, ticks, and leeches.

Many people have disappeared in the great swamp and you understand why if you've ever been inside its living black belly. Once you're inside, you're lost. Your compass may tell you from what direction you entered, but when you try to exit by the same route, you discover that the swamp has erased your path.

If you're an experienced explorer, you may have written down that you entered at 275 degrees northwest. But when you return along the same bearing in reverse, you keep going further and further, until you are miles beyond the point where you thought you'd come out. Turn back and things worsen. It's as if the swamp itself is an enormous carnivore which has just captured you for a snack.

The swamp's name says much about its nature. Okeefenokee is a Seminole Native American word meaning "trembling earth". And indeed, you can step on what seems to be solid ground and see trees quivering a hundred yards away. Every step you take may mean stepping into a bottomless pit of saturated rotting vegetation from which you'll never escape. The swamp will swallow you alive.

Mister Rabin, our thumbless scoutmaster, was still our leader because we had begged him to graduate with us into the Explorer Troop. He would go with us into the swamp, along with two other adult men whose primary desire was to fish the famous waters of Billy's Lake. We were to camp on Billy's Island.

I suppose it's an island but none of us ever found the opposite shore from the one on which we camped. The island itself was more swamp than dry land, and we had no canoes to cross the swamped areas. The ranger had rented us four flat-bottomed skiffs but balked at our request for the lighter, portable canoes. He was a wise man. We were safe so long as he knew exactly where we were and we couldn't enter the maze of waterways in which we certainly would have become lost.

In single file, we rowed our four boats bearing 12 campers down the canal which led to Billy's Lake and its island. The rangers kept this one canal free of dead logs but nevertheless we twice had to get out of the boats, enter the canal up to our waists, and bulldoze our way through thick patches of water weeds, sliding our boats over the surface. The slimy bottom of tangled dead branches and sunken logs grasped at our feet like hundreds of little hands. Unlike the rangers, who waded barefoot while cleaning the canal, I wore my combat boots.

When we were able to get back in our boats, Roundtree found a leech attached to his calf. He stared at it in disbelief then let out a cry that must have shaken the hearts of the wildest creatures in the swamp. Grabbing his hunting knife, we were about to cut it off when Mister Rabin stopped him, calmed him down, then placed a burning cigarette end to the leech. It dropped off and Roundtree overturned the boat trying to stomp the life out of it. We sagely hauled our butts back in and took our places without a word. Roundtree also sat back down, trembling and wide-eyed, while we were stripping and checking ourselves for other leeches.

We were chased away from our first choice for a campsite by a furious and bold black racer. The snake is remarkable because, though harmless, he is one of only two snakes in the world which will deliberately attack a human being. Every time we tried to put down our gear, the snake would rush at one of us with head high, its jaws open and obviously not bluffing. Its horrible hiss alone was enough to unnerve Willy, who ran back to his boat and refused to get back out.

Several of the others tried to kill it with sticks, which it easily dodged by diving into the undergrowth. Then he was after them again the second they turned their backs. I admired this proud serpent and its fearless heart.

I was supposed to be the specialist in snakes, as the others knew that I was fascinated by reptiles and kept in my bedroom a seven-foot long boa constrictor, an immature anaconda, a young timber rattlesnake, a four-foot South American tegu lizard, a five-foot iguana, two sizable alligators, and a variety of native harmless snakes. I had also delivered lectures about snakes to the student body, using live examples.

Honor bound, I tried to catch the racer. If I tried to grab it, however, it would dash into the bushes and literally glide through the branches beyond my reach. But when I walked away, it would rush me again and again, nipping at my heels. If I faced it, it dashed away again. I tried one time to push my way into the bushes so I could grab its tail. I did it, too. But the snake turned and lunged at my face with its mouth wide open. I let out a short, shrill shriek and fled like a scared rabbit.

"Hey, snakeman!" whooped Gooch, "Way to go!" The others also thought my screech was funny and their sarcasm angered me. But they fell silent when I dared any one of them to grab the serpent's tail as I had done. Ultimately we all obeyed the snake and returned to the boats. The racer escorted us back, snapping at our feet, making us scramble and trip. Willy was watching from his boat. His eyes looked like boiled eggs. He was pop-eyed and froggy faced. The racer was nothing compared to our next encounter. We saw many water moccasins and water snakes basking on cypress roots and logs but we weren't worried about them. However, when an alligator longer than the boat surfaced alongside Pootaroot's skiff, everyone instinctively froze in fear. The 'gators in the swamp grow to great size and become fearless because they aren't hunted.

What we failed to realize was that the 'gator was begging us for food and had no intention of attacking us. It bumped Pootaroot's boat with its head several times and we heard ol' Poot a'praying, "Oh, God! Oh, God! Jesus, help us!" Either God heard him or the 'gator grew bored, for he sank beneath the surface and we saw no more of him. But poor Pootaroot was shaken badly. His was the lead boat and he immediately headed for shore. That was our camping place, he announced, and he was going no further in the water.

The adults checked the area and it was sufficiently dry and open for a camp, so set us to work. Soon our tents were up, our Troop flag flying high from a makeshift pole, and a huge fire burning away. The adults had brought folding chairs with them – the big babies – but we

Explorers found and carried to the camp a giant dead log which we trimmed and used as a bench.

I wanted to set up my own camp away from the others but Mister Rabin refused to talk about it. Instead of a tent, I had a jungle hammock which had to be strung between two trees. For those of you who don't know, a jungle hammock is combination sleeping bag and tent, with mosquito netting between the top and bottom. Once hung, a camper can zip himself inside and sleep high off the ground without being pestered by insects.

I used this hammock as an excuse to separate myself from the others, claiming that the only trees spaced far enough apart were about thirty yards away from the main camp. Reluctantly, Mister Rabin let me string my hammock there. Placing his thumbless monkey's paw on my shoulder, he asked, "You ain't afraid of sleeping out there by yourself?"

"No, sir," I answered truthfully. There was much more to fear from Gooch, Willy, Roundtree, and Pootaroot, who might decide it funny to fill my hair with toothpaste or fill my canteen with kerosene during the night. Or they might throw a Cherry Bomb under my hammock or untie the ropes suspending me. I'd seen them do similar things to other sleeping boys. Thirty yards was really too close.

Besides, back then -- when I was all of fifteen years old -- I believed I possessed a sensitivity which the others did not share. To me, sitting by a campfire in the wild was a time to think

things over, be quiet and reverent, maybe read some Lord Byron, Wordsworth or Thoreau while listening to Nature's nighttime symphony. It was a time for a calm respite from humanity.
reason must not be expressed.

As Mister Rabin and I were talking, we heard a commotion in the main camp. Two raccoons had broken into one of the tents and were stealing our stores. They were brazen and we had to throw sticks at them to make them go away. They ambled off in no hurry, not bothering to look back at us, grinning at each other, smacking their jaws. When we checked the stores, we found they'd eaten or ruined half our wieners, several pounds of bacon, all of our cooking oil, and four dozen eggs.

It was an impressive theft as they had to unbuckle a number of complex straps to open the packs. They quite deserved their burglars' black masks. After that, we hung our supplies from a rope tied to a high branch, beyond the reach of the audacious fuzzy bandits. It was a good thing, too, because a black bear came to visit us too a few nights later.

Before dark, most of the others went fishing. Meanwhile, I went to my private little camp, built a fire, and began to look around. As luck would have it, there was a small wooden dock sticking a few yards off shore. I never told the others and they never found it. It was nearing dusk by then, and I decided that while the others slept, I would spend the night awake on the dock watching wildlife.

They returned just as darkness was falling

over the swamp. And I too returned. Though the middle of July, we felt cold and huddled around the fire. They'd caught twenty-two fish between them, ranging from huge ten-pound blackfish to four- or five-pound large-mouthed bass. They'd left them tied to a stringer submerged in the lake. (The next morning, they were to find that the mud and snapping turtles had left them nothing but grisly skeletons attached to heads with popping, terrified eyes. The fish had been eaten alive during the night.)

We'd been warned about the mosquitoes and now they came in their billions, swarming around us, flying down our throats, crawling down our ears, up our noses, getting in our eyes and, worst of all, bleeding us with thousands of tiny hypodermics at once. All but one of us had insect repellant and soon we were covered with greasy, stinking oil of eucalyptus.

But Willy, wanting to save money, had instead brought a half-gallon can of Black Flag insecticide. He stripped down to nothing and poured the poison all over himself, head to toe. And as we watched, very amused, Willy turned a blood red color. He looked like he'd been very badly burned but he claimed it was no worse than a sunburn and redressed himself.

It wasn't long before Willy, complaining of dizziness, walked off into the woods and vomited. Nevertheless, he came back smiling and sat again by the fire, full of jokes and wild stories. We watched him carefully. Whenever a mosquito landed on his skin, it dropped dead on the spot. "See there!" said Willy, "You guys wasted all that dough!"

About midnight, Willy began to complain about the heat, moving further and further down the log, away from the fire. The adults had already expressed concern and now all of us began to worry. Willy insisted he was fine, though his face was puffy, his eyes bloodshot, his nose running, his voice slurred, and his skin still the color of blood. He was also breaking out in saucer sized blisters.

We drew straws to decide who was going to take Willy back to the ranger station and on to an emergency room. One of the adults volunteered because he wanted to fish some more anyway. Roundtree and Pootaroot drew short pine needles and were thus chosen to go along. Loudly complaining and cursing Willy for his stupidity, Roundtree got up and stormed off towards the boats. But Pootaroot, who once had been such a terror to us, helped Willy up and held him around the shoulders as they hobbled together down to the shore.

Willy was weaker than water.

Mister Rabin told us to pack it in and hit the hay. We'd get up early for some fine fishing, he said. Don't worry about Willy, he said.

I returned to my private little camp as the others were crawling into their tents and rekindled my private little fire. Staring into the flames, I thought about this mighty swamp and its armored inhabitants. When the Spanish Conquistadors first saw the Okeefenokee, they sent reports back to Spain that the alligators were so numerous in the swamp that a person could walk across their backs – if one were so foolish to try.

Before the coming of the Europeans, adult alligators had no natural enemies and regularly grew to 25 or 30 feet long and as big around as four pregnant sows. Unlike mammals, they continue to grow throughout their lives and live a century or more unless killed by man or other 'gators. How I wished I could have seen those pristine days when dragons ruled! How the nights must have shuddered with their bellowing roars!

On impulse, I grabbed my flashlight and walked down to the little dock. An opossum was strolling across it as I approached and froze when blinded by my light. There was a full moon, so I turned it off and watched as the possum ambled away.

Lying belly down on the dock so I could gaze into the water, I was reminded of how I'd lain on a similar dock as a little child at Copperhead's Retreat. But instead of crayfish, frogs, minnows, and water snakes, this retreat was the home of antediluvian monsters who once ruled the world. I prayed an alligator would show himself.

Shining my flashlight into the water, I could see the shadows of catfish at least four feet long and crayfish the size of lobsters. Other shadows darted in and out of the light, fish as long as my arm but so swift they couldn't be seen. Around me, the swamp was in symphony, a million insects, frogs, and birds singing in eerie, primal harmony.

A yard long banded water snake swam into the light and lifted its head out of the water to see what was going on. He stared into the light for a few minutes, then quickly swam away. I heard something moving on the bank and, shining my light down, saw a moccasin sliding into the water.

Then he came. I saw a massive shape coming toward me under water. I knew at once it was an alligator and I began to tremble with excitement. I held the light as steady as I could and prepared myself not to panic. The dock was about ten inches over the water, which at that moment seemed far too close.

On he came and soon I saw him settle on the bottom. Even through the murky water, I could see his eyes glowing bright red in the beam from my flashlight. Very slowly, he began to rise to the surface without a movement, until at last his eyes, ears, and nostrils broke the surface. I found myself face to face with a creature whose ancestors had dwelled and prowled with dinosaurs.

His head was about two feet in length and one across. His body must have been about ten feet long. I could see his legs slowly paddling in the water, keeping him in this one place. I was about two yards away, nose to nose. Nothing moved but his legs as they paddled and his eyes as they considered whether I was edible.

I must confess that I was afraid. His mouth opened a few inches and I could see the dozens of teeth in that massive cavern. His nostrils opened and closed, as did the flaps over his ears. Several times he blinked, the nictitating membrane gleaming in the light.

Despite my fear, I wanted him to stay. I wanted to look into those red eyes until I could fathom what was happening inside that ancient mind, what was traveling through the primitive circuits of that dim reptilian brain. Was he thinking in any sense of the word? Inside that skull, were the ancestral memories of 300 million years of existence still recorded in some subliminal cerebral sphere?

Did he unconsciously miss those grand days when his ancestors swam with plesiosaurs in long lost inland seas?

I realized that he was hypnotized by the light and couldn't see me. I wanted him to see into me as I was trying to see into him. So I switched off the light, causing him to immediately sink to the bottom, only to rise again a few moments later. Again we stared into each other's eyes, but by the light of the moon instead of a flashlight.

He moved a little closer and time itself moved a little closer as I understood at last the living thread of life binding me to the alligator. We had come to be in this universe by the same process, animated by the same incomprehensible energy from a single divine current. We were both ephemeral apparitions, temporarily with form shaped by our environments. And ultimately we both would disappear as beings with mass, our living energy to perhaps continue and ultimately animate a new shape and form.

I realized the huge reptile was curious. There was life and some mysterious form of thought in those eyes. I was looking at a conscious being, no matter how different that consciousness was from my own. He sensed his existence even if he couldn't reflect upon it. He perceived sights, sounds, odors, warmth, cold, hunger, anger, desires, and touches. His emotions were as real and as sensitive as my own. I wasn't looking at a programmed mechanism but a conscious being with a mind of his own. And a questioning mind at that. He was as amazed by me as I was by him.

Eventually, he sank back into the depths and swam away. But I had seen into the past – 300 million years into the past – and even then living creatures were astonished by their own existence. From then on, I would see all living things as efforts of a living universe to witness itself.

Pootaroot, Roundtree and Willy were back from the emergency room when I returned to the main camp the next morning. The doctors and nurses had been flabbergasted, giving him not treatment but a thorough tongue-lashing. In the end, they told him he would likely get cancer and die. Then they gave him back to us.

He seemed a little worried that morning and his skin was still blistered and red. But he quickly cheered up and nothing bad happened during the next two weeks, except that he shed his skin like a snake.

Gooch openly scoffed at my story about the 'gator, calling me snakeman again. I was annoyed. I had read a page of Nature's book, in which he seemed to have no interest. This presumption caused me to sarcastically blurt out, "You don't care, do you? You've got your Big Sugar Daddy and cheerleaders and ball games and cars to think about. You're really an intelligent guy, Gooch head."

I was shocked by my own words. What had I done? This giant could kill me with one hand. . . his left hand! I expected a violent reaction and was preparing my apologies when Gooch turned his back, walked to the fire and sat down, staring into it with an expression of the deepest and worst kind of resignation.

Pootaroot gave me a scathing look, as if I had just killed a puppy. The others were staring first at me, then at Gooch. I was confused. Gooch was crying! It would have been easier to fight him and get the thrashing I had coming. Bewildered, I looked at Pootaroot for help. "You think you're so smart," he said; "You think you're smarter than everybody in the whole world."

It was impossible, I thought, that I'd hurt this rich boy's feelings. He was the enemy in my book. He had fancy clothes, a car, money, good looks. The girls were crazy about him and he was an ace athelete. He made all A's. He was a champion halfback on the football team. He'd lettered in his freshman year. And the worst thing about him, he was from a rich family. How could a commoner like me hurt the feelings of a pampered piss-ant prince?

Not knowing what else to do, I went over and sat next to him, saying I didn't mean what I'd said, I was only kidding. He turned so he was straddling the log bench and looked at me silently. I felt the blood rushing to my face. And as we sat there, staring at each other, it hit me that this was the first time in my life that I had looked into a person's eyes. Gooch couldn't have realized the lesson he was teaching me.

I saw in Gooch's eyes the same fire I had seen in the 'gator's eyes. Not anger, but a conscious light, a gleaming, radiant energy. I had bestowed intelligence and emotions on an alligator, then denied those same qualities in Gooch. The truth was in his face.

Every human being embodies the epitome of evolution on this planet. Each individual man or woman personifies humanity, the most complex and inexplicable of all creatures. We may be natural and integral to Nature's grand design, but we will never be able to fathom the infinite depths of our own human minds.

I could understand the 'gator but I could never really understand Gooch or any other human being. Individual people change constantly from one personality to another to another to another. They have as many facets as a diamond. It is impossible to peer within one without seeing a myriad of reflections from a thousand unknown mirrors of being.

After that morning, all of us were aware that rich kids are as complex and sensitive as ordinary folk. Gooch snapped out of his depression by noon and all of us feasted together on steak, eggs and potatoes, which we had wrapped in aluminum foil and thrown into the fire. But as we sat around the blaze swapping lies and drinking boiled coffee, it was not Gooch who felt alone but I.

I could never know what was inside my

friends and they would never tell me. Humanity was beyond my comprehension. As individuals or a species, they were as unknowable as the cosmos itself.

So after lunch I returned to my private camp, sat cross-legged by the fire, and read a few pages of a book I had brought along. But my mind was on other matters. I felt alone among my own kind. But I never felt alone when in the forests and swamps. The healing woods consoled me. The fire seemed a fine companion. There was nothing to worry about and all was well. The forest whispered soothing sounds into my ears, words which I couldn't interpret but could feel. The ancient swamps and forests were mankind's first home and this was where I wanted to be. At home.

Since that day, I have been aware of my ignorance of the human mind and heart. The 'gator has but one face -- the human being a hundred. Only a human can be consciously good and evil at the same time. Only a person can seem to be insensitive, yet be wounded by a mere word. Sometimes I still blurt out things I deeply regret later – but at least I'm aware that one facet of every human soul is forgiveness. And if one can find forgivbeness, every wound – no matter how deep – can be healed.

Chapter Four:
THE DAY OF THE BLACK BOAR

Wild boars were quite common in the Georgia woods in 1962. The local farmers, not willing to spend money on feed and not having enough slop left over from their tables, would release infant piglets into the woods in the fall, then hunt them down when killing time came round next year.

The little porkers prospered in the wild. In a month or two, they were large enough to have no natural enemies and continually rooted up the forest floor, devouring anything remotely resembling food – acorns, nuts, seeds, wild plants, bulbs, tubers, roots, bugs, grubs, maggots, salamanders, lizards, harmless snakes, copperheads, rattlesnakes, birds, eggs, rodents, baby woodchuck, baby rabbits, bay foxes, young skunk, opossum, raccoons, small dogs, fawn, newborn lamb, and even calves.

Within a year, the "little porkers" grew to over five feet long from tusk to tail, weighed between two and three hundred pounds, were strong as oxen, swift as deer, and short-tempered as wildcats. And they were eerily smart.

As for the farmers, they thought it great sport

to hunt the adults down from horseback at killing time, shooting them with large caliber handguns – mostly .44 caliber Ruger magnums. A few brave men used compound hunting bows and arrows. There was a very real element of danger in this bizarre and cruel harvest, for the feral swine had by then grown from their lower jaws four-inch long, razor-sharp tusks capable of disemboweling horse and hunter should they fall.

Most yearlings would squeal, run from the horses and get shot in the back. But a few wise old boars – those who had survived the previous year's hunt – would wait in thickets when hearing humans on horseback. There the boars would watch with tiny but wicked, intelligent eyes. There they waited with a terrifying patience. At the right moment, they would charge, deliberately slashing at the horseman's legs. Many good ol' Georgia boys back then boasted hideous scars where they had been gored and split open.

I thought sure I'd disturbed a black bear the first time I saw one of these brute feral pigs. I was exploring a foreboding cove called Pigeon Swamp outside Woodbury, looking for snakes, when it suddenly came crashing toward me, grunting and huffing, a giant black monstrous squealing mass of flashing, slobbering, snapping jaws.

So much younger and nimbler then than now, I scrambled up the tallest pine tree I could find and intertwined myself in the thinnest branches at the top, where I hoped no bear could follow. But it was no bear. I'd have been better off if it was.

Looking down at the crashing noise at the base of the tree, I saw its massive menacing body emerge from a thicket and stop dead in its tracks, peering around for me. It began tearing at the tree with its tusks when it found me hiding in its branches.

I had never seen such fury in an animal and found myself trembling so badly that the treetop threatened to break. I lowered myself to stronger branches, causing the boar to stop chopping its jaws and instead gaze at me with a peculiar, almost amused expression.

This particular boar was wise to the way of hunters and it could see I had no weapons. So it sat on its haunches like a dog and seemed to be thinking things over. I believe it was weighing its chances, trying to decide whether I was the ridiculous cowering coward I seemed to be or a real hunter with a concealed weapon.

I think the look in my eye gave him the information he wanted. He casually regained his legs, trotted stiff-legged and slowly around the tree, constantly watching me. Then with a profound contempt, he lifted his tail and defecated. This is what he thought of me. Then he sat again and waited. And waited. And waited.

As he watched me, other wild swine came silently out of the woods and joined the wait. They were all females, his harem, and would undoubtedly follow his example whatever happened. That they were females was of little comfort as their size was equal and their tusks only slightly smaller. As if to frighten me, every now and then one would chomp her jaws repeatedly, squealing and spewing a thick saliva.

I had been treed in the morning and the sun was getting hotter and hotter. My canteen held only a quart and I had four liver-loaf sandwiches in a little pack. When I began to grow desperate around noon, I threw one of the sandwiches to the beasts below. A wild scramble and the male had it, smacking his lips and looking up for more. Hoping to appease them, I threw down the remaining three. And indeed they seemed to calm.

About one o'clock, they suddenly got up as one pack and ambled off into the woods. I waited half an hour. No sound. Then a whole hour -- still only silence. With my heart in my throat, I began to descend. But when only a foot from the ground, the entire pack rushed back out of the woods with a horrid squealing and chopping of jaws, sending me scrambling back up the tree.

They were playing cat and mouse with me, taking out all the hatred they had generated for the humans who hunted them with deadly weapons from the safety of horses. Looking down, I imagined they were laughing. One flopped on its back, wiggling and jiggling its legs in apparent mirth. The dust flew. The other hogs grinned. But I did not see the humor of it.

I had plenty of time to think while pinned like that. For the first time in my life, I was the hunted – I was a treed, desperate soul wondering when or if I would escape. I've been told the pack would have scattered in panic if I had climbed down but I just don't believe it. No one who has been in that situation would believe it. The monsters were partying beneath me!

I have a hunter friend to whom I told this story and his response was scornful. "It was just a bunch of pigs," he said in a scathingly sarcastic voice, "and you were too chicken to chase them away."

"If that's true," I asked him, "then why did my grandfather lose three fingers to a hog? Why did my Uncle Joshua walk on crutches the rest of his life after being attacked by a boar? Why did my great great uncle write about the wild hogs eating wounded soldiers after battles of the Civil War?"

"You're just a coward, admit it!" and he gave a derisive, too loud laugh I'll never forget.

But he may be right, for in the end I escaped not by my wits but by losing them. As dusk neared and I became more and more afraid, I felt that fear turning into anger. I literally stopped thinking and allowed the beast inside to act.

From some part of myself usually hidden, I found that wild, untamed spark and allowed it to generate a current of madness which blinded all reason. Right before dark, I dropped to the ground and charged the boar with a furious bellow, swinging a branch like a club. The monster was astonished. After a moment's hesitation, he panicked and fled, followed by his squealing harem. I confess I wanted to kill him and would have had I been armed.

While walking home, wondering about the

loss of reason I had experienced, I remembered a similar incident that had happened the year before in the gymnasium of my high school.

His name was Peter Van Artsdalen, and he was awkward, skinny, smart, shy, bespectacled, and the butt of every bullying joke in the school. He wasn't different like everyone else is different; he was different in the wrong way. His knees would shake when he recited in class, and when he talked, his face would turn a beet red color. On the day I remembered, most of our class of 100 was in the gym watching our teams play basketball.

Most of us were in the bleachers. Peter was on the court and slipped but before he could rise a bully called Hambone – so called because he weighed one third of a ton and was shaped like a ham -- placed a foot on his chest. Whenever Peter tried to rise, Hambone would push him back down with a laugh. When he finally let Peter get up, he forced him into a corner of the gym and pushed him against the walls with his incredibly fat belly. Suddenly the whole class seemed to notice what was happening and rushed toward the torturer and torturee, forming a half circle around the show. Everyone was silent. All of us were fascinated, enjoying the perverse thrill which comes from watching other people suffer.

No one did anything, not even the coach, who simply watched with a sick smile on his face and arms crossed. Speaking only for myself, I wanted to help Peter but was afraid the bully and the crowd would turn on me if I brought attention to myself.

Almost all the other students, many of whom Peter had called his friends, watched with the same frighteningly cruel smirk on their lips that the coach had on his. The bully – this living Hambone – repeatedly shoved Peter against the wall and drew back his fist to see him cower. And the others laughed and laughed, especially the girls.

No one came to his defense. I would have tried but the fear of the mob turning on me was paralyzing. I knew I couldn't live with the humiliation and defeat Peter was doomed to live with. I think it was that precise, cold moment in time that I came to understand the appalling but instinctive pack mentality of the human race. They enjoyed watching Peter humiliated for the same reason wolves eat their wounded comrades. They were under the spell of the beast, rooting for the alpha male.

They were all enjoying themselves, those young men and women surrounding the torturer and his victim. I was dumbfounded by the mass betrayal -- I didn't realize the natural forces at work on them. The individual is reasonable as a rule, but the mob is irrational by that same rule. There is a thrill, a love of inflicting pain in others, lingering in the hearts of mankind . . . and in the hearts of pigs.

Like me, Peter finally escaped from his persecutor by going mad. Hambone kept pushing him and the others kept laughing until I saw Peter's eyes glitter with a malignant sparkle. I knew his mind was whirling. I knew he'd been pushed to that frightening outer limit called murder.

Later he was to tell me that things began to glow with vibrant, odd colors and he heard an angelic chorus in his head. He thought he was going to faint when, from somewhere so deep inside him that not even he could fathom it, he found that spark which – even if it didn't produce the courage of a hero – at least allowed him to take one more step on the eternal path toward human dignity.

In those intellectual terms, I describe what in simpler terms would be called insanity. In the gym that day, facing big fat Hambone, surrounded by smirking enemies -- Peter snapped. He attacked in a blind, mindless fury without knowing what he was doing. He pounded Hambone's head with his fists and kicked him to the floor. Three hundred pounds sprawled like a hog's cadaver. Hambone howled like a stuck pig and cried out, "You ain't got no call to hit me!" Peter began kicking him as hard as he could, tears streaming from his eyes, curses pouring from his mouth. . . he kicked until he could kick no more. He kicked until fat Hambone bled from the lips and nose. Then he collapsed and lay crying with his arms over his head. He had won but his classmates had seen him cry. His shame was unbearable.

But I had learned that men have no conscience, no mercy, when in mobs. The dumbfounded crowd stared at Peter with amazement and gradually, one by one, walked away. Hambone rose to his feet and said again, "You had no call to hit me."

It was Peter's time to be dumbfounded. "I beat you," he answered, "you . . . you . . .

Goddamned son of a bitch!" Hambone backed up as if scalded. Wide-eyed, he called him a crybaby God hater and walked away. Back then, especially in Georgia, you never never never said God's name in a curse. Only a person bereft of his senses would say such a thing. For the first time, his classmates and the coach realized this boy was dangerously near the edge. The coach had to work up the nerve to punish him. Fifteen minutes after the fight, Peter was sent to the office for his fifteen licks with a wooden paddle -- which he took like a man.

Later on, classmate by classmate, Peter's friends said how sorry they were they hadn't helped him, hadn't tried to stop this disgraceful battle. For years afterward, Peter treated us as baboons not to be trusted. But now that adulthood and time has made him wiser and more forgiving, I hope he has realized we kids were grasped by an ancient instinct which long ago lost its usefulness.

This story of torture, violence, and madness has a strange ending which even I can't understand. One year after that day of the black boar, I had my chance to hunt feral hogs from horseback. Instead of a handgun, I carried a borrowed U.S. Army M-1, filled with ammunition I had chosen especially for this job. Its soft lead tips were designed to mushroom in the animal's body, blowing a hole the size of a grapefruit in its flesh. I was after revenge.

My moment came right at dawn in those same Georgia woods and I flushed a black boar which might well have been my tormentor. I drew my sights along his spine. It would have been an easy kill. But something – I like to think some remnant of compassion – kept me from firing. With the tables reversed, I would not be the pig.

Later that night, as I sat alone by my campfire, I pondered this odd feeling. The other men in the hunting party were about fifty yards away. I had chosen to be alone. I looked at those hunters and I wondered who were the humans and who were the hogs. I wanted to believe that compassion is what separates man and beast but I knew it wasn't true. So after a few hours, I doused my fire and joined my fellow men by their fire. I knew I had to accept my role in the unchangeable drama of Nature.

Like it or not, I was a human being among human beings. It would take much work and time to rise above that terrible curse. Still, I was glad I had not killed my enemy when he was defenseless . . . though I knew he would have killed me without hesitation. That night I decided mankind's only chance was to overcome the bestial instincts tyrannizing our minds and hearts.

But I failed to consider Nature's prime rule: many humans are produced so a few may evolve into something higher than human. Out of the billions of human beings, I was later to discover, only a very few would have the strength and will to overcome their animal instincts. Only a very few would ever experience that exhilarating lifting of the spirit which comes with an evolved consciousness – with mind and heart enlightened enough to rise above the inevitably tragic processes at work on earth.

Today I believe this more than ever. If we are to evolve into higher, more spiritual beings, we must eventually and at whatever cost keep the beasts in our hearts forever imprisoned.

Chapter Five
<u>The Goat Man</u>

It was the first week of June when we met the Goat Man. The year was 1963 and through the years I had explored further and further into Georgia's Great Pine Woods. Sometimes I would spend all day hiking to my destination – if I had a destination – and camp wherever I found myself at dusk, hiking back the next day.

I was lucky then and had Jack Keefer as my constant companion, having decided between us to be best friends till death. This was a new and wonderful experience for me because I'd always felt like the outsider, friendless and unwanted by my peers. Jack changed all this for a few years – not till death, but till his family moved to another state. But for those few years we were inseparable. I loved him in a way that I have been unable to love anyone since. We were brothers – initiating ourselves into mutual commitment to each other by cutting our fingers and mixing our blood like the blood brothers we were.

We were both sixteen and both possessed by those remarkably clear ambitions of adolescence, those lofty dreams we knew would come true. We followed our destinies with the certainty of sleepwalkers. We especially liked to camp out on Pine Mountain, hiking all day and sitting around the campfire all night -- telling corny jokes and swapping our most private secrets about the girls we adored and the futures we envisioned.

So it was no surprise to our parents when Keefer and I decided that day in June to spend a whole week in the woods, as far as we could walk, all the way to the Flint River fifteen miles away. Though our mothers loudly objected and conspired to stop us, we left anyway at dawn on Monday morning. We carried our packs on our backs, our Stevens Savage .22 rifles in our hands, swinging casually by our sides.

So on we pushed forward, mile after mile, until at last we were in the secondary forest with its flat open floor of pine needles, and there we fell on the ground and panted and wheezed and rested till we could go further into the unspoiled woods.

We eventually picked ourselves up and pushed on for many more miles until at last we were stopped by the Flint River's shore and decided there to set up camp before the sun fell into that emerald sea of trees. We found a clearing full of grass and set up our tent and dug a shallow depression about three feet across and lined it with large stones for our fire. We got out our lines and sinkers and hooks, impaled victims from our box of worms, and tied the lines to sticks stuck along the shore and threw the baited hooks into the river.

During the night, catfish might strike. And with the morning sun we'd roast them on the rocks around the fire as the coffee boiled in our old copper-bottomed tin pot and potatoes baked in the coals.

That night we spent catching crayfish in shallows, which we carried back to the camp and boiled in the coffee pot until they turned a bright blood red and we could peel and feast on them like shrimp. The moon was full, a silver dollar in the sky, and beneath it we excitedly discussed the coming war, tanks and battleships, guns, hunting knives, crossbows and pistols, and how we would forever be rugged outdoorsmen, rejecting the glass and concrete world of humanity.

How painful to remember back to those uncluttered days when the future seemed bright as the moon. Now, of course, the moon is just as bright, but the future has become the past and in retrospect seems black as an eclipse. When Keefer left for Vietnam, to be shot through the leg and come limping back, I was at sea on the USS Raleigh and never saw him again.

But that night by the campfire, listening to the hushed whispers of the woods and river, under a sky resplendent with moon and a billion stars, we both understood with our flesh and blood that happiness was found in the wild.

Unable to sleep, we pulled a fallen log to the fire as a seat and knew we would be up all night. The frogs and spring peepers and a million trilling insects sung to us a harmony of unrequited love.

Down by the shore, a logging road paralleled the river. It hadn't been used in many years, for young pines grew in its track, and the rains had left deep valleys in its clay foundation. It was on this road that, around three in the morning, we heard a clamor of clanking tin cans, the groan of a wooden cart pulled by a sighing mule, and the barking of a dog.

We had heard tales of the Goat Man and never doubted his existence. But we never expected to meet this mythical man and his phantom beasts, no more than we expected to meet the wild Choctaw Indians who were rumored to still persist in the deepest Georgia wild.

When we first heard the commotion, we grabbed our rifles and fled to the woods to hide among the pines on our bellies to await this fantastic apparition. I was scared and Keefer was scared but we breathed as one and felt inside a delicious exhilaration.

In the light of the moon, we saw him rumbling and clanking up the road towards our camp. Our fire still burned and there was no way the Goat Man could miss seeing it. Surely he would stop and warm himself and his animal friends, seeking just as surely hot coffee and talk.

Local legend had it that the Goat Man was a mad nomad roaming the woods with dozens of goats and a shepherd dog. It was said he had once taught college and had a doctor's degree, but when his wife died of a dread disease had taken to the woods with his beasts to rove and never return to the towns. It was also said he was kind and wise, never failing to leave those he encountered feeling cleaner and better and free.

We believed because our algebra teacher,

Mr. McMurray, had said he'd met the Goat Man himself while fishing on the Chattahoochee River. He had told us how the Goat Man had spoke of the river as if he, too, wanted to flow swiftly to the sea, cleansing himself as he passed over the earth. Mr. McMurray was not the type to lie; he lacked the imagination to make up such a tale. And now, sure enough, the legend was coming toward our camp.

We saw him long before he arrived, riding an ancient four-wheeled wooden wagon pulled by a single mule. The wagon was filled with a thousand things, no doubt those survival tools and ammunition and wild plants and dried deer jerky needed to live in the woods. And perhaps he hauled, too, some treasures from his past which he couldn't bear to give up. Perhaps, deep in that wagon, were material memories of his lost wife.

Tied to the wagon with short ropes were about a dozen goats, who walked slowly beside the wagon, gazing around like resigned children on the way to a boring class at school. Running circles around them, occasionally nipping at their heels when they resisted the pull of the ropes, was a Shetland sheepdog with missing tail, all white but for black patches on his back and face. Later we would see that the dog had one blue eye and one brown eye.

The Goat Man himself was slumped in the seat, loosely holding the reigns, whistling some sad old song to himself.

The dog caught scent of us and began to bark crazily in the night, causing the goats to get nervous and begin trotting, trying to pull the old mule faster. The Goat Man perked up, peered at our fire, and pulled to a halt. He gazed at our camp for an endless five or ten minutes, before dismounting and walking towards it. His gait was slow but determined, as if he were on his way to the funeral of his brother, tormented but strong, grief-stricken but wise. He carried a rifle – a 1914 German Army 8mm Mauser, as we later learned – and a huge U.S. Marine survival knife strapped to his leg.

He was dressed in an old army uniform with a blanket draped over his back. He had on combat boots and from his web belt hung dozens of knotted cords. Around his neck was a khaki scarf, thrown carelessly over his shoulder.

But it was his face which mesmerized us. The second we saw that face, we both got up and headed toward the campfire to greet him. There was no madness or danger in those eyes, just a deep, resigned, unendurable melancholy. His expression was one that spoke of suffering, helplessness, anguish, and a terrible hard-won wisdom. Later in life I was to see that same expression on the faces of my friends returning from the jungle war, recovering from drug addictions, fighting nervous conditions, or trying to heal wounds left by constant loss.

He may have been thirty; he may have been seventy. His face was a tanned road map with a thousand etched lines. His eyes were blue and his hair streaked with gray. His hands were large, powerful, and very rough, with nails like clam shells, ridged and thick. His voice was soft, but deep and resonant, like that eerie ocean's surf you hear in a conch shell, soothing, elemental, timeless, primeval, and free.

"You boys from around here?" he asked.

Self-consciously we told him we were from Manchester, about seventy miles south of Atlanta, and we were camping out and looking for snakes to catch and take home. He thought about this information a minute or two, then said, "Why don't you leave them alone?"

"Oh, we're not going to kill them!" I said too excitedly, "We keep them as pets." I was intimidated that he had criticized us so quickly. But at the same time, I wanted desperately to know this strange man of the pine woods. It was as if I was peering into my own future if I failed at my goals.

He smiled, and to my surprise his teeth were perfect and white as a porcelain china cup. I was even more surprised, even shocked, when I saw the firelight reflected from stainless steel that held those teeth and formed most of his upper palate and part of his lower jaw.

But his diction and grammar were perfect, cultured, cultivated, and clearly annunciated, betraying at once this man's high intellect and sophisticated background. His voice seemed alien to his appearance, this man who lived in the wild with a bunch of goats.

Keefer's eyes were wide-open and perfectly round, with all the white showing around the irises, as he stared into the Goat Man's face. His mouth hung stupidly open. But the stranger either didn't notice or didn't care and invited himself to join us by the fire. I poured coffee into a canteen's cup and handed it to him.

The three of us talked a while about unimportant things until Keefer and I calmed down and relaxed beside this strangest of all strangers. As the tension left, the subject of our conversation became more serious, though we dared not ask any personal questions. He spoke of comets and meteors and exploding stars, while pointing out constellations in the night sky.

But Keefer couldn't control himself and blurted out, "Is it true you used to be a college professor?"

The Goat Man was silent. After a few moments, he admitted it as if guilty of murder. "Yes. A long time ago I taught zoology. But that time doesn't exist anymore. This moment is all that's real. This moment and the stars. You won't understand that for a long time."

There was no bitterness in his voice, but Keefer and I were old enough to realize we had touched a bruise on this man's troubled heart. I asked his name, and he said he was once called Paul. "But you can call me the Goat Man," he added with an odd little smile; "I guess your next question is why I have all these goats."

We confessed that we were wondering. He told us that he liked goats because they were clean and silent. And as he said clean, he stressed the word so much that we knew he meant much more than clinical cleanliness. He meant some kind of nobility which transcended the slovenly soul of mankind.

Keefer and I looked at the animals as they stood motionless around the wagon, at attention and watching us with aristocratic eyes. They were quiet, still, unafraid, patient. They were stately animals with straight legs and backs, immaculate coats, and lifted heads with chins high, as if proud and full of an unsullied integrity.

For an instant, I thought of St. Francis' sermons to the animals, but the image dissolved as quickly as it had come. This man didn't seem religious, just powerful, wise, tormented, and independent. Even then I saw the parallel between him and the goats. They lived together, but not as a herd. Each one was a lonesome, dignified, isolated, aloof individual, asking for nothing from men.

And like the Goat Man, their eyes revealed just how different they were from other creatures. Their pupils were vertical slits with blue irises. Some might call them demonic eyes, but I saw them only as clues to their essential separation from other beasts of the fields and forests.

The sheepdog worked up the courage to come and sit by the fire with us humans. He stared wistfully into the fire. He listened carefully as we talked, twisting his head curiously this way and that, as if trying to understand. His stump of a tail wiggled furiously whenever any of us would reach over and pat its grizzled old head. I knew that this was the Goat Man's friend, not the goats. The goats were a reminder of some sorrowful penitence he was paying to a God we had yet to know.

"Do you think you'll ever go home?" I asked

him, watching for that glitter from his stainless steel jaws.

"I am home," he replied; "You boys can understand that, can't you?"

"You like Nature better than people, right?" Keefer asked.

"Oh, people are Nature too," the Goat Man said; "It's just that people crowd together too much. Crowding squeezes people like me out. But they're Nature all right, people are. They're the mind of Nature. Nature thinks through our brains. Unfortunately, Nature has no heart."

He chuckled, knowing that we didn't know if he were serious or pulling our legs. He knew we didn't know what he was talking about. But I understand now what I couldn't then. Keefer picked up his rifle, gave a big grin and asked him, "Do you like guns? I love 'em. I'll bet you've killed some big bucks out here. Hell, you could blow a bear's head off with that cannon you got. You ever shoot a bear? You ever shoot a human being?"

I felt the same instincts as Keefer back then and shared his love of weapons and the thrill of the hunt, the lure of war. But I intuitively realized it might be best not to question this nomad about such things. In the firelight, I could see the remnants of some old but hideous wound around his mouth, despite the moustache and beard he wore.

The Goat Man showed only sadness as he said in a low tone, "I've killed to stay alive, but I don't enjoy it anymore. Sometimes I'm afraid I've got life all wrong and I'll have to answer for all the killing one day. That's why I learned to dry meat into jerky. One deer will last me all winter. A 'coon will last me two or three weeks, a beaver six."

Keefer evidently had not noticed the scars or the steel mouth and blatantly asked, "Were you ever in a war? That's where we're going as soon as we graduate. To Vietnam! That should be a real adventure!"

The Goat Man looked first at Keefer, then at me, then gazed again into the skies. "It'll be an adventure, all right," he told us; "It'll be something you'll never forget, no matter how hard you try. You may even bring home a reminder, like I did." He smiled wide, and for the first time Keefer saw that metal mouth and angry scars. He blushed, mumbling an apology, saying he didn't mean to offend anyone.

"That's okay, son, I'm not ashamed of it. I used to think it was my badge of honor, but now I don't think of it at all, except during the winter when it aches. Want to know how I got it?"

We did. We loved all war stories.

"I won't bore you with a long story. I was standing guard at Bastogne when the Germans launched their last big offensive in WWII. They came in wave after wave, wearing snow white uniforms. We were overwhelmed before we knew what hit us. A shell exploded as I was running for cover, and it knocked me out. When I woke up, there were American bodies strewn all over the place and squads of German soldiers were finishing off the survivors.

Anyone they heard groaning was bayoneted in the heart. Those who were silent got their teeth bashed in with a rifle butt to make sure they were dead. When they came by me, I stayed silent because to breathe too loud was to die. A soldier kicked me and bashed in my mouth, but I was silent and they moved on. And that's how I wound up with a steel and ivory mouth."

He looked at us to see if he had made an impact. He had. We were speechless at the horror of his tale. There could be no doubt it was true.

"Is that why you live out here all alone?" I asked.

"Oh, no, that's just life. Out of a thousand men, only I survived. I'm lucky in that respect. The Army fixed me up, gave me a new mouth and teeth, sent me to the university. I got married, taught college kids and for a long time was content. It wasn't the Germans to blame, it was just Nature."

I thought of the queen water snakes and tadpoles at Copperhead's Retreat, how they all had been killed but a few. Those who escaped had done so randomly. It was sheer chance that the Goat Man had survived.

Keefer seemed thunder struck. "Then why live out in the woods if you were happy? I don't understand. You might as well stay in the city."

The Goat Man thought about this, then seemed to choose his words very carefully as he answered, "Well, it's like this. What's the difference where I live if man is just an animal anyway? I mean, people may be smart and all that, but when you boil it down, they're just Nature and nothing more or less.

"That's the way things should be, of course, and the killing and brutality of some people is good for Nature. It thins out the species. But I just can't accept it, which means I'm inhuman. A freak of nature. I don't belong with other people. See what I'm getting at?"

We both confessed that we hadn't an idea of what he meant.

He sipped coffee, sighed, and looked at the dog. I wondered why he was bothering with us. What difference did it make if we understood him or not? Nevertheless, he continued, spurred on by some invisible force of which only he was aware.

"I've thought about this for twenty-five years, boys, and I think I've got it figured out. Nature gave us an instinct to kill -- we love it. We especially love to kill each other. The reason is to destroy the weak, so that only the strong are left. The strongest will reproduce and the result will be more and stronger young. That's the way things are and we ought to accept it. But I can't. Does that make sense?"

A little, we admitted, not at all sure of ourselves.

"I believe in God," the Goat Man said, "God is Nature. Everything is Nature." He looked into the sky, adding, "The earth we're on, everything on it, everything in the sky and the sky itself is Nature. It's all one living Nature."

Keefer was blushing in the firelight because he couldn't understand what he was hearing. But I was beginning to comprehend.

"It's only right, the way I feel about it," I said; "Lots of animals and people have to die so a few can survive. This is a hard world. Only the strongest deserve to live."

The Goat Man gazed into my eyes until I felt embarrassed and looked into the fire. He still wasn't angry, though both Keefer and I had shown only how thick-skulled we were.

"That's the first law of Nature, all right," the Goat Man agreed; "And that's where I made my big mistake. I cherished something too weak to live.

"I wanted compassion and conscience to evolve in Nature. I wanted the weak to live, too. I thought mankind was the mind of Nature and if enough of us with compassion survived, then we might evolve into the conscience of nature. But you're right, boy, death's as natural as the Smokey Mountains. It's me who's unnatural. I wanted the soft and beautiful to survive, not just the strongest.

"I tried to be above Nature's law and look what happened. I'm the Goat Man. I'm the goat. The joke's on me. This is what I get for trying to be more than just a man. Maybe in a thousand years Nature will grow a conscience. Maybe a few special people will survive out of the billions born, to evolve into a type of cosmic compassion."

He suddenly seemed very tired and got up, heading back toward his wagon and goats.

Before leaving us, though, he added, "You boys enjoy yourselves while you're still strong. Just let me give you one piece of advice. Don't ever care for anyone too weak to live. She'll . . ." and here he stopped, seemed confused for a moment, then corrected himself.

"I mean, it will die young for sure and you'll be left alone. You're smarter to care only for the very strong. It'll survive if it's smart and strong enough. Until that thousand years has passed, you're better off without a conscience. Don't ever love the weak. I wish I never had."

It was to be many years before I realized he was talking about his dead wife. She was the weak one he had wanted to live, only to watch helplessly as Nature's ruthless laws were enforced.

And with that last cryptic message, he walked back to his wagon, clambered aboard, and patted his mule on the neck. He rolled away in a racket of metal and hoofs and creaking wooden wheels. I thought we'd said the wrong things to him, but he stopped, turned and waved goodbye. And then the Goat Man rolled into the darkness and was gone.

An hour later and the sun came up. We pulled in and cooked our catfish, speaking in hushed tones about the night. Neither of us had learned anything. Later, Keefer and I were both to sign up in the military and volunteer to fight in Vietnam. Keefer would be wounded at Da Nang. I would spend those years on a warship.

I don't know if Keefer ever killed a human being but I expect he did, as I've heard he was in many lethal firefights. Me? I don't know if I have or not.

When the USA invaded the Dominican Republic in 1965, my ship helped shell the beaches. I was part of a "pom pom" gun crew, and we fired three-inch diameter, high-explosive shells point blank into the tops of the palm trees lining the shore. When the firing stopped at last, through my field glasses I could see bodies hanging from those now leafless palms.

Snipers had tied themselves to the trunks, unwittingly making themselves sitting ducks. I enjoyed it then and hoped we'd blown away every one of the native defenders. Now I pray we didn't hit any at all. I pray it was the other warships and other cannon crews who killed those people.

I wonder how Keefer feels now about the Goat Man's talk of mankind evolving into the "consciousness and compassion of Nature"? As for me, I've had enough killing and seen enough dying. I wish I could find the Goat Man again now that his words make sense to me.

I would tell him there is hope yet. We are not snakes and frogs. Those few of us who can survive may indeed eventually infuse a trace of benevolent compassion into Nature's otherwise merciless heart.

Chapter Six
The Bat With a Human Face

One of the most sad but most exciting possibilities about exploring the woods of the deep South is the discovery of an abandoned cotton plantation. You know immediately what it is by the tall white columns still standing in front, usually supporting the remains of an elegant veranda where Southern Belles once sipped mint tea and entertained their dashing young men in gray.

Behind these columns, the great mansion itself is most often on the ground or caved in, ringed behind with the remains of smaller one-room houses which served as slave quarters just over a century ago. You know you're gazing at a culture based on grace which was wiped off the face of the earth in its prime. Sadly, you also realize that such grandeur was made possible only by slavery, that shame of the South, which demanded its destruction.

The ruined mansion I found in early July of 1963 had been abandoned so long ago that it was completely covered with blossoming honeysuckle vines, wisteria, and thorny rose bushes, all descendants of what must have been in 1860 a splendid, carefully tended garden. Where the cotton fields had been 125 years ago now stood a forest of loblolly pines.

Such places are dangerous, of course, and you have to take every step with caution, less you stab yourself on a nail, trip on rubble, or fall into an ancient root cellar. And, most dangerous, you might fall into the plantation's abandoned well, which is almost always around the ruins nearby and covered with deceptively solid vegetation or rotting wood. Most have been filled with debris; but others are still open.

This was a thrilling event for me because I had heard legends about desperate plantation owners throwing gold, silver, jewelry, coins, and weapons into their wells to keep family wealth out of the bloody hands of Yankee looters.

I found the old well underneath the remains of the old well house, right where it would be expected to be found. Lifting the roof aside, I peered down into a deep black hole which seemed to pierce right into the earth's dark millennial heart. By dropping stones, I could tell it was about 75 feet deep but practically dry.

There was nothing more I could do, so I rushed home to prepare for the next day's exploration into the dark hole. The following morning, I arrived with 200 feet of stout rope, 500 feet of thin but strong nylon cord, two six-volt, waterproof flashlights with an interchangeable red filter, a small collapsible spade, and a police whistle. That was my standard spelunker's kit.

I tied one end of the rope around a strong tree and dropped the other end into the well. I had knotted the rope every yard to make it easier to climb. Carefully, I descended to the bottom to see what I could find, my heart beating wildly. The thought of treasure was forever the epitome of my hope. But I found nothing on the bottom but a small stream of water inhabited by a few frogs.

Looking up, the large hole I had entered now looked like a full silver moon suspended in a starless night. It reminded me of the Chinese concept of "well vision", with which a human being could see only a fraction of existence but remained ignorant of all other reality. Those who believed they understood existence from that tiny glimpse of light were fools, their minds swollen with pretense and self-delusion.

So there I was, buried in my own delusion, pompous in my teenager's well-vision of the world. The well was dry, and there was no treasure; therefore, there was nothing at all. I was to be proven wrong.

It slowly occurred to me that the tiny stream had to come from somewhere and go somewhere. To my left, the water seemed to seep from the wall of the well. But on my right, the stream emptied into a chamber about the size of a pumpkin. Squeezing through, I found myself in a small muddy chamber which emptied through yet another small opening. I managed to get through it also, with a little help from my spade.

But there I discovered a chamber so huge my flashlight would hardly reach the opposite wall. It was about the size of a double-wide mobile home, maybe 12 feet wide and 50 feet long. The trickle of water turned into a fair-size stream about four to six feet across, so clear I could see what looked like jewels scattered on the bottom. A close look at some of these 'jewels' revealed they were polished quartz crystals.

Dozens of massive stalactites enameled with folding skirts of limestone hung from the ceiling, gleaming and dripping, glimmering with prismatic colors. Many were larger than me, joining thick, ringed stalagmites rising up from the floor to form a stone forest of unearthly hourglass pillars. Its roof and walls, though slimy and wet, nevertheless shimmered as if made of eerily translucent and shimmering green glass, a frozen crystalline waterfall welded to the chamber's surface.

There was no sign of treasure, so I crossed the chamber, getting caked in mud from head to toe because in many spots I had to crawl on my belly to slither under ledges or climb over giant slabs of slippery rock blocking the path. I was exhausted by the time I had crossed it – mostly on my belly like an alligator – and I drank deeply from the stream when I finally reached the other end.

It tasted of lime and time, more time than I could imagine, enough time to create a thousand earths and evolve a billion creatures. The tiny stream had carved this hollow intestine through the fossilized gut of the planet, turning it into this resplendent green cathedral.

There was no outlet on the other side, so I backtracked and found one leading off to one side where the stream forked. At this point, I tied my nylon cord around a stalagmite and let it trail behind me. But I didn't need it, for as I entered the next chamber, I saw a shadowy myriad of shapes fluttering around. I heard a shrill chattering filling the air and smelled an intense, awful odor of methane gas and ammonia. I had crawled into a colony of bats.

This chamber was a little larger, about the size of a large living room, and there were hundreds of bats. The floor, instead of being muddy, was covered with bat droppings. Oh, I was startled, all right! But having studied bats in their own territory before, I knew they were harmless and wouldn't touch me even if I turned the light off. Their echo-location systems (sonar) were so perfected that they could "see" me as well in the dark as in the light. Probably better, as the flashlight must have been painful to their eyes.

I was filthy anyway, so I stuck my arm into the dung pile and was amazed to find I couldn't find the bottom. The colony must have been there hundreds, maybe thousands, of years. Hearing a swishing sound, I looked just in time to see a large black snake trying to escape the light. It probably visited the bat chamber regularly to feed off the old, injured, and very young which occasionally fell from their roost on the ceiling.

I turned off the light and lay there a few minutes contemplating what I had seen and where I was. These were creatures of darkness, yet made of atoms comprised of radiation, of light, heat, the energy of the stars. They cast off no light of their own, but if I had been able to see it, each bat would gleam with the pulsating energy from which they were created. If I could but see it, the black snake would look like a ribbon of light.

As soon as the light was out, the colony settled down except for a continuous chirping, no doubt in anger at my intrusion and a request for me to leave. I could feel the moisture seeping through my clothes.

I thought too that the basic element of life was present in the stream. All living things are comprised mostly of water, carrying as it were the aquatic environment where they were first spawned with them in their waterproof skins. And what is water? It too is made of atoms cooked in the hearts of stars, a vital component of all flesh, the very basis of all existence and evolution. It is liquified light, providing the malleable substance of ever-changing forms of life.

It was at this point that the acrid, overpowering smell of ammonia forced my mind off cosmology and back to this flesh and blood reality. I was getting cold, and my eyes were watering. A man's mind can escape the earth, but his body is forever chained to it.

I have never been afraid of the creepy-crawly creatures which seem to frighten most people; rather, man is the animal I most fear. So I was in no hurry to leave this secret colony, despite my discomfort.

When looking for wildlife on the surface at night, I had discovered that a red filter over my flashlight seemed to be invisible to most animals, while allowing me to watch their actions. That's why I had brought one with me into the well, so I could see in the darkness without disturbing any interesting life forms I may find inside.

Slipping it over the lens of my flashlight, I pointed the red light up at the surface. Though the colony had quieted down, most of them seemed to be looking at me with their mouths open. Actually, they were making sure I remained harmless by keeping their sonar fixed on my person. They apparently did not see the red light.

Their devilish faces were truly demonic apparitions. But I knew their expressions were evolutionary adaptations which enabled them to hear the sonic and subsonic echoes of the cries they continually emitted through their noses. The horrific expressions were necessary, and only mankind judged them as evil because of their appearance.

I studied the colony carefully, noticing that the clever little mammals lived in a highly structured society, just like most other gregarious species, including mankind. Along the edges were the large dominant males guarding their harems, which consisted of as many females they could guard at one time. The other males, who had not been strong enough to win mates or territory, lived in clusters alongside the harems, careful never to cross the border into the big males' territory.

To my surprise, there appeared what seemed to be a "day-care" center, in which many immature bats were watched and cared for by one or two "nurse" bats. To top this off, the bats shared food with each other, like man and bees, a phenomenon simply not seen in the animal kingdom except among bats.

Because of my former writings, I will not repeat myself here except to comment that we seriously underestimate the intelligence of bats. They are devoted parents, loyal to their neighbors, share food amongst themselves, keep themselves scrupulously clean, groom each other, show affection to each other, and seem to have a communication system as yet beyond our comprehension.

So it was with great interest I watched the colony. But several yards from the colony's border, I noticed what seemed to be a white cocoon of some insect. Moving to within a food of it, I saw it was a white bat wrapped completely in its wings. I had never seen an albino bat, so gently tapped it with a finger. Instantly it unfolded its wings and glared at me with fangs barred and making little screaming sounds.

At these sounds, the rest of the colony dropped from their upside down perches and again filled the air, a seemingly mass confusion but with no collisions. Could the white bat have been a sentinel? The answer proved to be yes and no.

Peering as closely as possible at the white bat, I was shocked to see it lacked the devilish facial expression of the others. The ears were only half the normal size and hairless, with translucent skin. There were no facial appendages, and its face was strangely flat, like an ape's or human's face. But what amazed me most were the eyes, which were three times bigger than the tiny eyes of the others and a pale blue in color. True albinos have red eyes, so this had to be a freak of nature.

But as I stared into that humanesque face, I was moved in an inexpressible way. She had two breasts like a human and a worried expression, with wrinkles over her brow. Her face was that of a concrete angel. As the colony began to settle down again as she stopped screaming, she suddenly made a break for cover and scampered upside down across the ceiling toward the safety of the colony.

But as she reached the edge of the bat town, the regular residents began chattering and snapping their jaws at her. When she tried to physically break into the community, she was beaten off by vicious swats from the insiders' wings. Dropping off the ceiling, she circled around in the air a few minutes, then returned to where she had been when I'd first noticed her. Landing about a yard from my head, she walked back to her original roost only a few inches from my eyes.

To the devil-faced bats, she was the monster. To me, her face held a wonderful and unexpected beauty. She jumped toward me several times, flapping her wings and snapping her jaws and chattering in that shrill, wild cry. I realized she was ready to sacrifice her life to save the colony, fighting me with only the bravery in her heart and delicate membranoid wings, even though the colony had rejected her.

I wanted to calm her, letting her know I was harmless. I also wanted to capture her and keep her safe in a darkened cage with plenty to eat, warmth, and nothing to fear. And that's when I made the mistake.

Not being able to frighten me off, she hung upside down again and wrapped herself once more in those white wings. She had given up, and no one from the neighboring village of bats would help her. She must have felt at that moment absolutely abandoned and doomed to her fate.

So I very gently cupped my hand around her and was gong to slowly enclose her in my hand without harming her.

But at my touch, she went limp, and before I could think, was slowly fluttering like an oak leaf to the floor. No sooner had she hit bottom than the black snake had her by the head. She flapped her wings a few times, giving the snake the appearance of a flying serpent, then gave up and disappeared forever down the serpent's throat.

I did not blame the snake, who was only disposing of the dead. I did not blame the colony, which was only protecting the integrity of its genetic pool. Of course I did not blame her. If there was a villain, it could only be me, though I had acted out of mercy and sympathy for an outcast.

It was Nature's laws at work again, against which I could not fight. The thought that she was made of light and was now free did not lessen my guilt nor sadness. A rare and beautiful and selfless creature had died at my mere touch. Just as she could not rise above the instinct to love and protect the colony, I could not rise above my human belief in compassion and justice. I had learned to rise above the instinct to kill, but in the end Nature had her merciless way just the same.

As the years have passed, I have seen the same thing happen in school, on board ship, in barracks, in college, and at work. The outsider is isolated from the crowd, ignored, and rejected, no matter how hard the outcast tries to fit in. Eventually the outcast is left alone and vulnerable, prey to a thousand dangers.

It is Nature, and only a complete lifting of the human spirit can beat that law of Odd Man Out. When we can accept the odd, the strange, the sick and injured . . . then we can begin our ascent to that higher consciousness which I pray is our destiny.

Chapter Seven
The Goldenrod Spider

The next story I must tell is not about my human friends or even the woods or its inhabitants. Instead, it's about a little spider – one I came to know intimately during the spring, summer and fall of 1964.

The blue-eyed bat had gotten to me and my best friend Jack Keefer had moved to Fort Lauderdale in Florida. The bat I did not miss but I felt inconsolably lost and saddened by the loss of Jack. Willy and the others were still around, of course, but I had unfortunately narrowed my closest affections to Jack. Since his absence, I'd become even more introverted and alone than before. I was not only a loner in my mind, but a loner in fact. It had been forced upon me – one day Jack was with me and then he was gone, good as dead.

Fears had long been embedded in my conscious and unconscious minds that I was doomed to be an outsider, forever alone, no friends, no wife, no family, no one at all. I felt I was the shunned bat at the edge of society, unwilling to leave but unable to belong. When in town, I was hanging by my heels in a forbidden realm inhabited only by humans. Only in the wild was I at home.

So I sought solace in the great pine forest, as

I always had – but I also sought diversion from my so painful and unhappy realizations about myself. I was a Goat Man already, my heart and mind closed to other people. I was damned and doomed to wander the world like the eternal wanderer who insulted Christ. I had insulted humanity by simply existing and filling space. And in turn they had insulted me. My adolescent heart was bitter.

I needed to walk much and think much and wander through the pine woods alone. And alone I wandered and wandered, till one day I came across a patch of goldenrods. Upon one creature here, I thought, I will focus my attention and keep a diary in its name. A diary of a bug. That seemed appropriate, for I was not enough in this world to be a bat or an alligator – certainly not a human being! At best, I was a bug, infinitesimal and invisible and absolutely alone.

I chose one goldenrod to examine and under one of its blossoms I found what appeared to be a crab. It was in a way – the creature was a goldenrod spider, better known to most as a crab spider. She was waiting beneath the petals, so I waited too, watching this lonesome bug as it lived its apparently pointless and boring life. It was a hot day and I needed a drink, so tipped my canteen and downed some tea I carried along. I wondered if she – the spider – was thirsty.

She must have been, for a honeybee came to the flower for nectar – only to be ambushed by my spider friend. She drank her victim's blood thirstily before dropping its body to the ground. I decided then and there to watch this wondrous eight-legged creature throughout her life to see whether indeed her existence was brave or cowardly, magic or tragic, productive or pointless, dramatic or banal. So every few days I would hike to her little goldenrod home and watch her drama unfold – at the same time reading all I could find in books about her kind. I saw much and I read much . . . .

The story of the goldenrod crab spider (Thomisidae) is a natural drama filled with beautiful temptresses, treachery, throat-cutting, vampirism, bizarre romance, cruelly discarded lovers, motherly devotion, and ultimate sacrifice. We might even say with pity that the lady crab spider's life is human, all too human.

She belongs to the order Arachnidae, which covers this earth like dead leaves in fall. She and her kind apparently evolved from marine creatures which invaded the land eons ago to prey upon the earlier evolved insects. The first spider known to paleontologists is the Palaeotenzia Crassipes, which was discovered in fossilized peat from the Devonian Age about 370 million years ago in Aberdeen, Scotland.

Since that beginning, the spiders have become one of the most numerous creatures in existence. They're comprised of over one million family members and 25,000 varieties, which can survive almost anywhere, from 22,000 feet above sea level on Mt. Everest to two miles beneath the land in coal mines.

A biologically sound forest has from 50 to 150 spiders for every square yard. One acre of field grass in late summer will average 2.25 million spiders. If the spiders living on that acre were to spin one continous thread for ten days, it would stretch from here to the moon. And in the United States, the amount of insects consumed by spiders per year would weigh more than the total weight of all human beings living there.

Having gossiped about her ancestry and plentiful kinfolk, we can now return to the drama of the lady crab spider and her all too human life.

We pick up her story in spring when she has blossomed with the flowers into an extraordinarily beautiful but cunning young woman spider. She is one of the loveliest creatures in the lush gardens and fields, dressed in bright colors to match the petals underneath which she hides with ambush on her killer's mind.

She is for a while the gem of all spiders. Her soft humanesque skin is either milky white or lemon yellow. On her legs are pink bracelets, and on her back is a deep red or purple arabesque of intertwined scrollwork. A pale green ribbon runs along both sides of her breast. And her movements are as fluid and rhythmic as a ballerina.

A 17th Century minister and spider expert named Reverend E. Topsell put it in these words: "The skin is so soft, smooth, polished and neat that she precedes the softest skinned maidens and the daintiest and most beautiful strumpets. . . . she has fingers that the most gallant virgins desire to have theirs like them, long, slender, round and of exact feeling. There is no man nor creature that can compare with her."

How can such a seductive lady be called a crab? Because from a distance she resembles the crustacean, and when she walks, she walks sideways.

Unlike many spiders, she can't spin nets or snares to capture her victims. Nevertheless, she deserves the name "spider," which derived from the Old English words "spannan" (to spin) and "spyther" (a spinner). Though she doesn't use her silk to trap her supper, she does use it to create truly luxurious homes for her many children.

Besides, she hardly needs traps when she is such an accomplished and refined bushwacker.

She spends her spring days hiding beneath flowers, waiting with a terrifying patience for her favorite prey, the honeybee. She is passionately addicted to bee blood, so when one approaches the blossom, tastes its intoxicating nectar and becomes befloured with yellow pollen, the crab spider craftily circles behind it, creeps silently forward, and when in range rushes the poor bee and stabs it in the neck with scientific precision.

Her hollow fangs easily penetrate the unarmored neck, injecting a venom whose potency is at least ten times greater than any rattlesnake's.

The neurotoxic venom acts within one second to paralyze the ambushed victim, who has time for only one harmless sting randomly delivered into the air. The neck is the most vulnerable spot because it is there where the cervical nerve center is housed. The attacker holds on as the bee's legs stiffen and its sting can sting no more.

The murderess sucks her still living victim's blood on the spot. And when her vampirish deed is done, she scornfully flings the drained corpse aside. The presence of a crab spider can be deducted by the pile of bee bodies scattered callously on the ground beneath her home.

But perhaps we are wrong to judge too quickly this back stabber and cutter of bees' throats. As the great entomologist J. Henry Fabre reminds us in his classic studies, "We are all ogres, men and beasts alike. . . the main thing is that the morsel be tender and savory."

Our stomachs indeed determine the morality of our actions. (And who but mankind determines the morality of spiders and ogres?)

As it says in the fairy tales, however, the ogre eats the children of others but loves her own. When it comes to her own family, there is no better mother than the otherwise devious crab spider. The cruel vampire becomes a model of devotion where her own children are concerned.

But she has no respect or love for her husband, who comes sneaking to her like a thief in the night when mating time comes in May. This may have something to do with the uniquely unromantic manner of their affair.

The male crab spider has all reproductive organs encased in his abdomen. When the urge to mate sounds its irresistible music, the male seeks and finds a female. Before mating, however, he has to work himself up to a crescendo of lust by stimulating his own testes.

Upon ejaculation, the male deposits the drops of sperm into a tiny basket he has woven for that purpose. This bag is then held in his two sensitive, handlike limbs situated in front of the two front legs (pedipaps).

Like a beggar, the male offers this bag of sperm to the female. If she's in the mood and doesn't have a headache, she accepts the present and inserts it into her own genital openings (she has three), where the bag splits and releases the sperm to fertilize the eggs.

She seems to resent this unmanly way of making love and immediately ignores the poor wretch of a husband as if he didn't exist. If she's in a bad mood, she may kill and eat him. But this is of little consequence, as the male would die in a few hours at any rate. Better that his substance should help nourish the mother, who will live another five to six weeks while guarding her unhatched brood of about 1,000 spiderling eggs.

Like some human females, she seems glad to have the juicy biological process of fertilization out of the way. Now she can devote herself exclusively to building a nursery chamber and keeping watch over it.

The aging lady crab spider stops eating and will eventually starve to death because she can't spare time from guarding her eggs. She climbs to the top of the rose or other bush, where she finds a shriveled and curled leaf among the twigs. This she encloses in a bag of silk to keep it from falling or opening, making a little cottage for her babies. On the inside, she spins layer after layer of snow-white, soft-as-down silk. In the end, she has a cozy cone as soft inside as the inside of a mammal's womb.

Last, she lays her egg sack inside the artificial womb and seals it shut with silk. On top of this structure, she adds a canopy of silk to provide herself with a home of her own. Backing into the canopy, she is prepared to fight to the death to protect her young. She won't leave the canopy again. Eventually, it will become her coffin.

As the weeks pass, she begins to lose her beauty as her skin becomes wrinkled and shriveled. Her colors fade; her figure becomes gaunt and thin. She becomes an old hag, so to speak, with a nasty temper and a one-track mind. Again, very much like certain human females I have known. If an intruder appears or an innocent passerby brushes the nest, she hurries from her watch tower and threatens the interloper by waving her front legs.

If you tease her with a straw, she will fight the weapon with her fists, parrying like a boxer about to deliver a knockout punch. If you knock her out of her canopy to the ground, she will climb right back up and fight you again. It makes no difference that you outweigh her several thousand times.

If you use a kinder approach and offer her a living bee, she will ignore it. She has absolutely no interest in anything other than hatching her children.

By the time the great day arrives five or six weeks later in July, she has shriveled into a literal shell of her formerly beautiful self. When she feels the restless babies squirming to be released after hatching, she uses her last drop of energy to tear a hole in the nest's door, despite her shattered health.

And as the spiderlings escape, she returns to her canopy to gently let herself die, hugging the nest and turning into a shriveled relic.

But Nature never really dies. The mother spider and even her poor excuse of a husband live on in their many children, who are about to perform one of the most spectacular and wonderful exodus in the natural kingdom. Thousands of the babies, replicas of their parents, gather on top of the nest and begin spinning a network of crisscrossing rope bridges.

Though spiders operate instinctively and not rationally, the baby spiderlings seem very human as they appear to play and cavort on the bridges. They often fall off, saving themselves with safety lines, then climbing back up. They bumble about clumsily, as if discovering their legs and what can be done with them. They hesitate, retreat, and seem to think things over as they explore the new world which opened with the egg nest.

When the sun warms them and the breeze swings them back and forth, the thousand plus spiderlings spin long threads which the breeze pulls aloft. The spiderlings climb these ropes like Indian fakirs, until the breeze snaps them from their moorings and carry them into the skies.

Anyone lucky enough to observe this mass ascension into the heavens would be astonished to see the tiny spiderlings sparkling in the sun like a fireworks display exploding overhead.

Flaming in the sun like so many gleaming points of light, the little spiders are the sparks of that living firework display. The tiny animals shine and glimmer as they soar into space. It's a sight not easily forgotten and returns to the memory as one more proof of the grand scheme of life happening here on earth.

The spiderling parachutists are sometimes carried miles upward and thousands of miles over the land. Ships hundreds of miles at sea have been known to be landing sites for baby crab spiders. They are literally blown to the four corners of the world.

And when they finally descend, they begin anew the cycle leading to the crab spiders' vampirish lifestyle, loveless reproduction, selfless sacrifice, gentle death, and wonderful rebirth as sparkling airborne riders of the wind.

Chapter Eight
Fear in the Desert Night

Not everyone knows there are deserts in Georgia. I know only because I have found them in the midst of giant forested areas and canebrakes. The sand is as pure and bleached white as that found on the seashore. All that grows on it are prickly pear cacti; almost leafless, thorny bushes; and miserable little tufts of grass. Most of those I've found were only a few acres in size, but further south one can find them more than a half-mile across.

In these unexpected biospheres live a remarkable variety of animals, including countless insects, scorpions, spiders, toads, four or five species of lizards, plenty of snakes, box turtles, gopher tortoises, songbirds, field mice, cottontail rabbits, foxes, and an occasional bobcat.

This surprising variety is because the deserts are so small that its inhabitants can find ample water in the surrounding woods or canebrakes. And as a rule, it's much cooler – though I have been in more than one when the temperature hit well over one hundred degrees Fahrenheit.

The particular day I'm thinking about was in July, 1964, and I had spent the morning watching my spider friend in the goldenrod patch. By noon, however, I grew hungry for new adventures and moved deeper into the pine woods. After an hour or so, I found myself wandering around in the canebrakes bordering Big Liza Creek as she flows through Woodland in Meriwether County.

I'd not found nor seen anything alive except a few six-lined race runner lizards who shot across the sand like rockets. But I wasn't disappointed, for I'd found a perfect bird-point arrowhead made of flint and was feeling very pleased with myself.

The edge of the desert appeared suddenly, as inside the canebrake I could see only a few feet ahead. But then, there it was, an expanse of white sand dotted with cacti and shriveled bushes.

It's believed these barren sand dunes resulted from cotton plantations which were cultivated there a century ago. The story is that the cotton farmers didn't know enough agriculture to rotate their crops, so their cotton patches – used year after year – ultimately lost the nutrients in the soil and refused to grow anything at all. Without a vegetative cover, it ultimately lost its topsoil.

Other theories range from leaking herbicide tanks to asteroids striking the earth to strip mines filled with sand. But my favorite – if not the most probable – explanation is that these deserts occur where human beings have launched snake eradication hunts. With the snakes gone, rodents and insects quickly over-populated and ate every bit of vegetation on the site.

Ironically, the humans had to abandon the fruitless land, giving it back to the snakes who returned in droves to reoccupy the area and its oversupply of insects, toads, lizards, mice, rats, and rabbits.

In any event, I found myself that hot day on the edge of a miniature Sahara. Dropping my pack and taking off my army canvas belt – holding a folding spade, canteen, compass, first aid kit, .22 caliber revolver, and survival knife – I walked into the desert to see what I could find.

All the way across, about a quarter mile, I could see what looked like the remains of a house. As I got closer, I saw that it had been made of pinewood slabs now warped and gray, a tin roof, a well, an out-house, and a rusty old children's swing with broken chains hanging forlornly from the top bar. I felt an odd, inconsolable sadness as I neared the shack. One room had collapsed, along with the roof of the porch. And a wooden rocking chair rocked slowly in the breeze as if someone had deserted it at my approach.

I mentioned a breeze. It was the only breeze in the whole desert and existed only because three huge oaks grew in the front yard, each at least 80 feet tall, with thick branches towering over the old homestead like clouds. Their presence caused an updraft of air into the tops, where it was cooled and came settling back to earth in gentle breezes. Maybe some wise old carpenter had planned it to be a natural air-conditioner for his wife and children before his land died.

Inside I found the remains of furniture: bed springs, a black pot-bellied wood-burning stove, a chest of drawers, and a shattered mirror strewn all over the floor. Old yellowed letters were everywhere, sad remnants of a life spent vainly trying to stay in contact with old friends and family. In what must have been the living room was an intact fireplace, a rotting carpet, and a damp, dilapidated couch which smelled strongly of mouse urine and worse.

I casually glanced at some of the letters; they spoke of the weather, births, catastrophes, diseases, crops, crop failures, church events . . . mundane things. But outside I found something that spoke volumes of what this family had endured. A cast iron fence, twenty-foot square and partly fallen over, surrounded five tombstones half buried in the sand.

By digging around one of them, I could read, "Here lies Sarah Jackson, 1938-1942" and four others with different first names and dates. There lay a grandfather, a grandmother, a wife who had lived 93 years, a husband who had expired after 65 years, and three children, two of whom hadn't lived over the age of sixteen. A third daughter, Mary, had lived to see her fiftieth sunset in 1950.

Far to one side, outside the fence, was a sixth tombstone. It was hard to read, so I poured water over its dusty face until I could read "Zeke Parker, Sgt., U. S. Army, Killed Honorably, 1944". I wondered by Zeke had been buried outside the family plot. Why had his body been returned here from Europe?

Returning to the house, I began to read the letters' return addresses. I eventually found about a dozen with Zeke's name on them, all addressed to Mary, the eldest daughter. I was embarrassed to be prying into what must have been a tragic past for them. I read the letters, nevertheless.

I'm trying to recall events which happened over forty years ago, so I can only approximate what was in his letters. Other than the usual romantic laments and promises, there were short descriptions of what he had seen and felt while fighting in the muddy WWII foxholes of Italy.

"Mary, it's not the days which frighten me. During the daylight, battles come and go and I might be terrified for a few minutes. But that's a different thing than the fear I speak of. While the sun is shining, I'm a killer and I confess I love it. Just last week I charged a German .30 caliber machine gun nest, killing not only all four crewmen but their dog, too. Oh ho ho! We can't have any Nazi puppies, now can we?

"Once I crawled under a Panzer PKW4 German tank to attach a bomb. It tried to crush me by spinning on one track, but I escaped, and it went up in a ball of fire. Imagine, me a hero! How those Jerries burn, all lit up like Roman candles, covered with black oil and red fire, trailing dirty smoke and yelling 'Freund! Freund!' They gave me a medal for that and I laughed.

"I know this isn't the Ezekiel Parker you knew in Columbus. It's the sunlight that does these things to my mind. It's as if the sunlight itself is a powerful drug. I feel invincible and so damn brave I know I'm going to get killed if I don't stop playing with these German devils.

"But – and these letters must be kept absolutely secret – at night I get scared. Even in the dugout with the other men, I lay trembling and sweating and waiting for the enemy to come to kill us with our boots off, still in our bunks. And being on guard duty is even worse, for it's there I become a craven, cowering, crying coward. Only you can I say these things to, Mary. I cry like a baby lost in the night.

"The sun does something to me, but the moon does something else. At night I cringe in my foxhole, constantly looking around, trying to pierce that darkness. I can feel the enemy's eyes on me, watching and waiting for me to relax just one moment, so he can thrust that jagged German bayonet through my chest. Or even worse, the moon makes me imagine being shot in the head. Suddenly the lights go out and that's it. The sun, the moon. Oh, Mary, you can't possibly understand!

"In the sunlight I look at the dead and feel like a conqueror. But at night, when I look at them, I feel only a hideous fear that tomorrow I will also be lying cold and stiff and bloated in that no man's land between the enemy and I."

Suddenly I felt uneasy and dropped the letter to the floor. I could only guess what happened to Zeke and Mary. No doubt Zeke was killed before they could marry, so maybe the family had refused to let his remains rest peacefully in the family plot. Perhaps he had no other family. Or maybe Mary's mother or father read the letter I had and discovered his fear of the night. And on those grounds refused to allow a "coward" to be buried in the family plot.

By then it was getting dark, and I had to

pitch camp somewhere. There was a waning moon as I recrossed the desert toward my gear. But when I got to where it should be, it wasn't there. I searched the area, with no results. I looked into the forest, and it seemed a far different woods than that from which I had emerged. It was sinister and seemed to absorb the moonlight like a sponge.

I searched until the moon had moved halfway across the sky. To go into the woods would mean to get lost. If I stayed where I was, at least I could circle the desert in the dawn till I found my gear. That's what I decided was best.

Like a dog in the snow, I scooped out a burrow in the sand and curled up inside, thinking to sleep. But – as many campouts as I had been on alone – this one was different. No fire. No light. No hammock, tent, or sleeping bag. No weapons. Just me, lying in a foxhole of my own with only the foreboding trees around. I tried everything I knew not to think of Zeke's letter. I found myself hoping he had been killed under the sun, not under the moon.

I lay there for an hour, cursing myself for leaving my gear behind because I was too lazy to carry it across the desert. Then I got up and walked back and forth, back and forth, pacing like a caged panther. There were no German snipers or sappers out there, but nevertheless I was less of a man in the dark than in the light. Zeke had stumbled across something which made me aware that a part of me was afraid of the night.

Had the first half-human, half-ape hominids huddled together at night, peering fearfully into the dark for marauding giant cats, cave bears, pythons, crocodiles? Had the moon watched as they prayed the sun would rise again? Did this fear ultimately become instinct, which still lingers in the darkest circuits of our brains, surfacing when we again find ourselves helpless in the gloomy nighttime black?

And was it the sun that transformed them every morning into savage, killer-ape hominids, banding together to boldly hunt the mammoth and the rhino and the hoofed beasts of the veldt? Had sunlight so filled them with aggression and courage that they ultimately covered and controlled the entire planet? And when fire was at last tamed, did its light free them from fear of the night and give them reign of all creation?

Some scientists claim the pineal gland in the center of our brains is the remnant of a third eye in our reptilian ancestors. The actual eye itself can still be seen and still functions on most lizards, who in fact "go mad" in the sunlight. In man, it is said, the sun still affects this gland, gathering currents of energy from the sun which gives to the body strength, bravery, and power. Left in the dark, this energy is no longer gathered, and we become again terrified primates, huddled together and pining away in our collective fear of the dark.

This was my first taste of the mystery that is light. Eventually I would come to see that we are affected by sunlight because we are made of sunlight. All matter is made of light and ultimately must return to light. This is modern physics and can be proved in laboratories and by the energy released when atoms are torn apart. What seems to be firm matter is actually energy held together by nuclear forces.

I hope this is true, for it means upon our deaths we are actualized as light beams and can shed this cumbersome clay called flesh. We can again become part of that universal light known by Christians and cannibals alike as God.

Virtually all religious texts, including the Bible, say without allegory that God is light. And darkness is death. If so, everything that exists – light, heat, energy, radiation, living flesh, stones, suns, and matter – is alive, forever suspended in a lifeless black ocean of dark. And we, being made of light, are thus made of life. Call it God, if you wish.

And when these clay shells finally crumble, releasing the light from which we are created, we glow brightly again, streak out into the endless skies, and take our places as resplendent angels in the heavenly consciousness of infinite light.

Our atoms are conceived in the hearts of stars. Perhaps our essential spirits of light are manifested over and over into material beings which seem to exist and then not to exist. Perhaps this is the grandest of illusions and willed for unknown reasons by that incomprehensible higher consciousness called God.

So maybe there is substance after all to those many legends which end with their heroes changing back to light upon their deaths, to spend eternity in the heavens as living, conscious suns.

Perhaps right now, even as I contemplate

our fates, there burns overhead a pensive new sun whose earthly name was Zeke.

Chapter Nine
The Horn Hunter

Graduation from Manchester High School came as a sock for all of us in June of 1965. Though I'd considered myself an outsider, I was nevertheless astonished to see my classmates packing their bags and leaving town for their respective futures. I had no idea what I wanted to do – my mother's philosophy had been to let each child choose their own paths.

All I knew was that I wanted to be a zoologist. But that seemed impossible at the time because I was so terrified of speaking in front of groups that I thought I'd never make it through college, even if I had the money.

So, with no reasonable plan in place, I hitch-hiked down to Fort Lauderdale to find my former best friend Jack Keefer. I did find him and for a few weeks slept on the beaches wrapped up in a blanket while he slept at his home. We'd get together during the days to scheme and make plans. I wanted to join the army and fight in Vietnam. He wanted to join the navy. We hit an impasse.

I finally changed my mind and the both of us went to the navy recruiter, took the tests, endured the physical, and prepared to ship out. But Jack backed out at the last moment, sending me alone to navy boot camp in Great Lakes, Illinois. From there I was assigned to the USS

Raleigh and began my stint in the military. I never saw Jack again.

After years at sea, I finally returned to Georgia in 1969. The war raged on and the people I'd known so intimately in my youth had changed. The division between them was a void so wide no bridge could cross it. It was a void so deep no one could understand what was at the bottom of it. Into this emotional war between generations and ideals I returned, at once confused and appalled.

In 1965 I had signed a pact with the government: I would serve on one of their warships if they would later pay my college costs. I kept my end of the bargain and they kept theirs. So in the spring of '69 I signed up at Columbus College in Columbus, Georgia, 70 miles from the town in which I'd lived before.

After years of living with 400 sailors and 1,200 marines on a ship, I wanted more than anything silence and solitude. But these elusive desires were not to be found at Columbus College, for it too was torn apart by angry differing opinions about the Asian war, racial equality, and dreadful urban riots. I took no sides, seeing myself as an experienced hardened veteran of war and peace who was now above the petty squabbling of children. I was 21 then and perhaps more absurd than I am now, but not as absurd as those who shouted opinions formed only from watching television and listening to protest music.

I was beginning to learn that wisdom is silence. I didn't hang around the hippies, nor did I hang around the squares (to use the idioms of the time). But I enjoyed the atmosphere of learning – for those who wanted to learn – and the classical arts and knowledge so far beyond the violence of the time.

To my delight, Pootaroot, Gooch, and Willy were also at the college. I was told Roundtree had bought the farm, and in my naivety believed he had really gone into farming. Eventually I was to learn he had died of gangrene after stepping on a Viet Cong punjee stick.

This was a piece of bamboo sharpened at both ends, one of which was dipped in a mixture of human excrement and rotting animal flesh. The other end was stuck into the ground along jungle paths, waiting for unsuspecting American soldiers to step on them. The punjee sticks were so sharp they penetrated even the tough soles of combat boots and pierced the unlucky foot inside. The result was inevitable blood poisoning and eventual gangrene. I was also told Roundtree had died whispering about God and a girl named Irene.

Pootaroot was studying pre-med and enrolled in ROTC; Marvin was already a senior majoring in zoology; Gooch was studying business; and Willy, that clown of our youth, was taking the easiest classes he could find, wanting only the free money from Uncle Sam. I began as a biology major but ultimately switched to literature because it, like the forests, offered an escape from the fractured human world around me.

Pootaroot, Gooch, and Willy had fought in the jungles of Viet Nam, and all bore scars, though not of the physical kind. The scars were deeper, embedded in their very hearts and souls. I felt guilty in a way for not going with them into the fire and fear of battle. But they all told me the same thing: I had done the wisest thing. I wasn't so sure when I looked into their faces. They had that expression I call "the dark eye", which you see only in the eyes of those men and women who have seen and known great suffering, yet survived. The prevalence of fear was forever branded on those faces.

I was astonished by the change in their appearance until someone told me I, too, looked different. And gazing at myself in a mirror, comparing the reflection to photos of myself in high school, I realized that I too had the dark eye. I too had been branded.

The war was over for all of us except Willy. He returned not as the Willy we had known, but a Willy who smoked marijuana, drank 100 proof vodka by the fifth, had been addicted to heroin, and who had contracted an unknown venereal disease which had his doctors baffled. He had been in a firefight or two, and had burned many villages. But mainly he had been intoxicated throughout his tour of combat duty in Southeast Asia. Now, back in America, he was still at war. A war not with Communists, but himself.

None of us were adolescents anymore, and all of us felt a terrible sadness at the loss of our innocence. Even in the Navy I had seen unspeakable atrocities and animalistic horrors which had stripped all humanity from the men who had taken part. Like my friends, I had war stories to tell but could not tell them without feeling as if an eagle had my throat clasped in its talons. So we were silent, though feeling a contempt for the younger students who were so terrified of the draft or so foolish they bragged of wanting to go to war.

We were not "the gang" anymore. We were four individuals who had been reshaped by violent experiences, turning us into cynical men who wanted only to regain our integrity and freedom. But it was too late for that. Forever we would be prisoners of our memories. To this day, we are prisoners.

Out of nostalgia, we still called each other by our nicknames. Mine had been shortened from "Snakeman" to simply "Snake", though Pootaroot often argued that I should be called turtle because I looked more like a turtle than a snake. But fortunately for me, his arguments failed, and "Snake" my moniker remained.

Marvin had not served his country and never visited us, maybe due to shame, maybe due to a wisdom he thought he possessed. Instead of the war, the four of us – without Marvin – talked about the present and those things which were distracting us from those bad years.

Pootaroot spoke of surgery, his great interest. Gooch spoke of the warehouse he had already leased and planned to use as a base to build a business empire. I spoke of the wildlife I wanted to study in the wild and the books I had been reading. But Willy had a different kind of current passion: he was a hunter, he told us, a Horn Hunter.

What he meant was that he had developed an obsession about hunting deer while overseas and now, finally, could indulge in his desire to fill his house with mounted deer heads he had killed. He wanted more than anything to hunt for magnificent buck deer and display their magnificent antlers. Though always a little tipsy from the vodka he kept in his car, his eyes would sparkle and but out when he spoke of hunting for "horns".

To him, we understood very quickly, it was the only way to rid himself of the rage consuming him alive from the inside out. Killing relieved him of a terrible hunger for revenge against the Viet Cong who had made him cower in fear and thrust him into his present unbearable life of alcohol addiction and incurable sickness.

To him, deer symbolized those little Asian soldiers hiding in the jungle, unseen and unpredictable. He had failed at hunting them down. Instead, the Viet Cong had conquered him and left him with a hatred of life and a deep-rooted need to kill. Willy, being the weakest willed among us, gave in completely to his instinct to destroy. And he induced a psychological transferal, changing his loathing of the Viet Cong to a loathing of wild animals.

I think he also needed to restore his self-respect and believed he could do so by proving himself a hunter. Those antlers he wanted on his walls would, he thought, bring him respect and admiration. He wanted to be and be known as the greatest of all Horn Hunters in Georgia.

The first hunting season came and went

without Willy even seeing a deer, much less killing one. But I had seen plenty of deer, for I had been camping out every weekend since I had enrolled in college. The woods were the only place I could find the peace I so desperately sought.

Week after week, I packed my gear and drove my Volkswagen to a place I knew in Talbot County, about fifty miles from Columbus. I parked next to a service station and hiked into the pine forest. The woods were thick, and few people entered them, not even during hunting season. I could set up my tent, build a fire, and ramble around in the forest looking for animals to watch. I could read my books without being laughed at, and I could think for hours on end without distractions.

I knew the area from my high school years. I found and could watch the wildlife because I had long ago discovered the secret of not disturbing the natural routine of their lives. It's simple, really. All you have to do is find a likely place, preferably by a creek, and lie down.

If you lie still long enough, the creatures around you will eventually lose their fear and resume their lives in the open. The birds will begin to sing again; squirrels will scamper across the forest floor; the snakes will prowl, and the little mammals resume their perpetual search for food.

Through binoculars, I could watch the deer which never came closer than 200 yards of me. They traveled in small herds consisting of one big dominant buck male, his harem of females, and often a few younger bucks. I learned through that first year that the experienced older bucks developed a cleverness and elusiveness far beyond anything expected of any hoofed creatures.

For example, when coming to a clearing, the old buck would allow the impetuous young bucks to enter first. If nothing happened, he sent the does to graze. Only after seeing for himself that the area was safe did he come out of hiding and join the others. He seemed to realize that he was the prime target of any human predators who may be waiting in ambush.

I saw the same intelligence at work when the herd had to cross the dirt road between them and the next grazing area. First the inexperienced young bucks were sent across. Then the does. And last, the big boss buck himself would sprint across, but only after looking both ways up and down the road for danger.

It was during my second year of observation that the tragedy struck. And I confess immediately that it was my lack of judgment which leads to calamity. Though I knew of several herds, one had particularly impressed itself into my imagination. It was a typical small herd with only five individuals, but lead by a giant buck with a long scar on its rump where he'd previously been shot but escaped and lived, that much wiser for his experience.

The first year, he had grown a huge rack of antlers which he used adroitly to drive off four different bucks who attempted to win his territory and harem through combat. The challengers were all driven away after a great clashing of horns as buck charged buck head on, their antlers clattering like swords as they collided and sparred with each other. Every time the interlopers, bleeding and exhausted, fled back into the woods.

The bullet-scarred buck was king and spent hours and hours sharpening his antlers by rubbing them against trees. When they dropped off after mating season, he seemed smaller and less majestic, but still retained his prominence among the deer population, traveling with suspicious eyes scanning everywhere around the herd, his chin held high and his nostrils constantly flaring and sniffing the air for dubious scents.

The second year something happened which made him even more imperial and impressive than before, but which also brought his doom. As his antlers were developing, some strange chemistry within made them sensitive to sunlight. By the time they had grown full size in late summer, they had been bleached snow white, gleaming like ivory.

That white rack spanned at least a yard across and boasted sixteen points. It was a spectacular sight, and I never failed to gasp in wonder when I saw him.

Unfortunately, when deer season opened that second year, I made the mistake of mentioning the fantastic buck to my friends at college. Willy was there, and his mouth fell open. He actually drooled as I described that wonderful set of snow-white antlers.

Willy began dogging me constantly, begging me to lead him to the Valley of Deer, as he called it. I refused over and over again, but eventually began to weaken as Willy became more and more desperate, more and more pathetic, in his begging.

"Listen," he told me, "I'm nothing in the world. I fought in Viet Nam, and I'm still nothing. I'm screwed up, you understand? You don't know what it's like, having everybody think you're a psycho, a coward. Listen, just between you and me. Maybe you'll understand. In Viet Nam, it was my nerves, you see? I cracked, I admit it. I had to be shipped back. But my nerves . . . you can't understand what it's like.

"You can save me, Snake. Listen, if I could bag a real trophy, a real set of horns, I'd be cured. It's eating me alive, thinking I'm a nut case, a coward, afraid of the woods, afraid of guns, afraid of everything!

"I got to prove it to myself that I ain't. When I was in the jungle, my patrol got ambushed by the Cong. I saw one of them hiding and aimed my rifle . . . but it was my nerves, it wasn't me . . . I just started shaking so hard I couldn't shoot. Listen, I just got to know the truth. I'm begging you, Snake, take me to this place you been going to. My life, my mind, is in your hands. I'm a drunk, a pot-head, but you can help me snap out of it. Where's your friendship? Where's your humanity? Please take me out there."

The man wept like a child, holding on to my hand. I was embarrassed beyond the telling of it. I didn't know what to say. Willy slumped like a rag doll, limp and crying. His problems were real enough. I weighed his future against the future of the buck with white antlers. The buck was old. He only had a few more years before he'd either be shot or defeated by a younger buck.

But Willy, my old friend. Maybe I could help him, as he claimed. Maybe I could make a difference, return his self-esteem and confidence. Maybe I could wipe away the stain of cowardice and relieve him from the guilt resulting from his failure of nerve in the jungle that day. After all, I might have folded too if ambushed like that. I decided I'd take him with me next time.

I was torn apart by doubt. But I was still young and found it impossible to be hard on a friend. If being a Horn Hunter would change things for him, who was I to refuse to help? After all, he was human, like me. Most of all, it was the weeping which got to me. I couldn't bear it.

So the first week of deer season found Willy and I at my camp in Talbot County. He had with him a Winchester 30:30 rifle, camouflaged with netting. Both of us were wearing camouflage head to foot, plus camouflaged hats and faces streaked with charcoal. We splattered turpentine over ourselves to remove our human scents, hoping turpentine would not smell strange in a pine forest.

Willy was nervous, and I was angry. At him, at myself, and at life itself for putting me in this position. I was convinced I had been forced to choose between an animal's life and the sanity of a friend. There was no choice in such a situation. I was in a vice.

Before dawn, we left for the hunt. There was

vodka on Willy's breath, making me uneasy because I was in front of his rifle. He was too eager to kill, I knew, and insisted he unload or at least uncock the gun. I couldn't help but feel a contempt for him, worsened by a contempt for myself.

Nevertheless, I did as I had promised. At the foot of a low ridge covered with underbrush and young pines, I told him we would have to crawl on our bellies – silently – to the top and peek over the summit into a valley below. That was where I most often saw the herd and the proud old buck.

I hoped they wouldn't be there, but they were. The old buck was drinking out of a stream but looked up just as we cleared the top. Both of us froze motionless until he decided all was safe and again drank deeply from the water. I could feel Willy trembling; the ground shook, and the pine needles rustled.

The herd was about 150 yards from us. Willy set his sights on the old buck's chest, just behind the shoulder blade. "This is it," said Willy, and fired. The shot broke the morning air like a bolt of lightning. All the herd panicked and were gone before the sound had died out. All except the buck with the snow-white horns, who did three backwards somersaults and fell to the earth with a thud, his legs twisted grotesquely beneath him.

"I got him! I got him!" Willy shouted, running toward the buck. I scrambled after him, shouting for him to be careful. But the buck didn't move. When we were about twenty yards away, I noticed the buck's eyes were closed and shouted a warning to Willy. "He's still alive!" I cried out;

"Don't get any closer!" But Willy ignored me and kept running.

When he was only ten or fifteen feet away, the old buck suddenly jumped to his feet and charged, his white antlers down and aimed at Willy's chest. Everything changed in an instant; now it was Willy's turn to panic. Dropping his rifle, he turned and ran, with the wounded buck right behind, furiously snorting and grunting, blood spouting out of his side. Willy was screaming.

But then the deer collapsed and dropped dead, his eyes wide open, blood pouring out of his mouth. Willy looked back, saw the buck down again, pulled out his hunting knife and ran back, jumping on the dead animal's back and slicing its throat, spewing blood all over himself. Willy was crying and shouting, "He's dead this time! He's dead this time!"

It was a full ten minutes before Willy stopped mutilating the carcass and found some semblance of composure. He was covered with blood head to foot. He licked some of it off his hands, smiling. He looked at me with an insane joy, still trembling with excitement. I only stood silently and watched, holding his rifle in my hands.

I let Willy gut the dead king, and together we drug the heavy corpse back to camp, where we strung him head down from a tree branch to let the remaining blood drain out. Willy was a happy man, dancing around his victim with a wild and irrepressible glee.

Later, after I snapped a few snapshots of him with the box camera he'd brought, Willy decapitated the king and held his head in the air, shouting to his malignant gods, "I killed him! I killed him!"

Indeed, he had killed the old buck. And indeed, Willy no longer seemed tortured by doubts of cowardice and "folding up". Later that year, the head with the magnificent white antlers hung in his den, looking as if it were still alive, with glass eyes that seemed to follow you around the room, staring accusingly and without pity. His friends came and saw the great buck. And Willy was right, they admired him for his hunting prowess. In his mind and theirs, Willy was somebody special.

A hunting magazine eventually heard about the unique trophy and featured a photo of Willy posing with the buck. In Willy's hand was his Winchester 30:30, and on his face was a big, satisfied grin.

But I grew more doubtful every year about the role I played in making Willy's reputation as a Horn Hunter. In fact, after twenty years had passed I was convinced that I had harmed Willy more than those Viet Cong who ambushed him that fateful day in the jungle.

Willy had become a fanatical hunter after I'd helped him make his first kill, hunting every animal when it came into season, from squirrels to wild duck to wild turkey to quail to fox to raccoon to bear to anything that lived in the forest or fields. During bow season, he had hunted with a bow and arrow. During rifle season, he had used a rifle. And during black powder season, he had used a musket. The walls in every room of his house were lined with mounted heads.

One room was reserved for animals with horns or antlers. There were not only a dozen or so deer, but a moose, several mountain goats, an antelope, and even a caribou. Willy had traveled far in his quest for horns. He had won dozens of gold cups and trophies for his talent as a Horn Hunter.

He had stopped drinking and stopped taking drugs shortly after I helped him sacrifice the buck with the gleaming white antlers. He had stopped attending psychotherapy. He certainly had never cried in my presence again. He believed I was one of his best friends and treated me to steak dinners whenever I was in town.

But I had realized one day in the fall – when the pine needles crunched beneath my boots as I wandered around in the woods – just what I had done to Willy. I remembered the nervous, weeping man I had tried to help become "a real man". I was so young then. I hadn't understood that his tears were already turning him into a real man. I had interfered with that metamorphosis.

The last time I visited was just last year, and he had a very strange story to tell me. It was hard to believe over 20 years had passed, and it seemed 1970 again. He was not the confident Horn Hunter anymore. His eyes were tired; the dark eye had returned. He again seemed nervous and unsure of himself. He was so shaky that his coffee cup rattled when he picked it up. His steak dinner grew cold as he stared miserably at the cooked meat. "I don't know if I can ever eat flesh again," he mumbled, pushing his plate aside.

"Something's wrong with me," he said;

"something bad. I keep waking up at night burning up and sweating, then I get cold and start shivering. It's weird, man. I don't understand it.

"Just last night I got scared in the night. I mean scared. Every time I shut my eyes, I'd see . . . well, like monsters. Sort of like big gorillas with thick dirty necks, all bellowing and coming after me. It was like a delirium or something."

I wondered if he had not returned to drinking, for alcohol can certainly arouse monsters in the dark hours of the night. But he insisted he hadn't returned to the bottle.

"I can't explain it right," he told me; "I was just lying there when I got panicky. I had to get up and walk around. You know, pacing like a cat in a cage. Nervous and scared. I went into the den and sat in my chair. I looked at that buck I killed with you that day, the one with the white antlers. Man, it stared back at me, and I was afraid. All those heads I'd cut off those animals were staring at me. I heard things, too. Whimpers and little crying noises. It was out of an Edgar Allen Poe story."

He was sweating and every now and then would suddenly gasp for a breath.

"I began thinking, what if I've got cancer or some disease? What if I'm gong crazy? Those heads . . . all those heads I'd severed from bodies. It's crazy. Man, I knew I had to pay for all those lives. I never got back to sleep. I had to go outside and walk. I even hoped the cops'd pick me up and take me to an emergency room or something. I'm going nuts."

I knew what he had felt, and I knew too well what was happening to him. After a while, when he'd calmed down a bit, I told him what I thought.

"Look, Willy," I said, "trust me. You're not going crazy. You're not sick. What's happening is your conscience has finally escaped. You're not dying, you're changing. You're becoming more than you were, and it hurts. The next time one of these panic attacks hits you, tell yourself it will end and you'll be a better man for it. Just stick it out. Your caterpillar days are over. Now it's time for you to grow wings and fly."

He looked at me with those whipped, tired, beaten eyes, and I knew he would be okay. The suffering I had helped him avoid had now returned. But I knew this time to let things take their natural course. Yes, he was going to be in pain for a while. He would be forced to examine himself and his values. And, like so many of us, he would discover a meaning to his existence, a state of grace, which until then had remained hidden within the brutish forces compelling him to kill.

The Horn Hunter was dying, and I did not mourn his passing. The person now forming in the chrysalis of time would be like the sun escaping from a terrible darkness. The light would soon burst out and illuminate the shadow which had covered Willy since the killing of the white-antlered buck.

Willy was beginning to understand at last.

Chapter Ten
<u>Fox and Hounds</u>

Passing through Maryland one night on my motorcyle in 1970, I stopped at a coffee house to relax and listen to some live jazz before continuing my journey back to Georgia. I'd been to Detroit to meet with a herpetological society, only to discover they required a bachelor's degree in zoology, which I lacked. I'd been treated with a scathing indifference and was in less than a good mood.

I sat at small round candle lit table and ordered black coffee and a blood & tongue Greek sausage. I was alone at the only empty table and wanted nothing else but to brood angrily to myself about the academic snobbery of American science.

The coffee house, inappropriately called "The Bottom of the Barrel," was a bit too glitzy and rich for me. The women wore strapless evening gowns and the men black ties and formal black jackets. I didn't much care about such things and possessed a bold rebel stance I've regretfully lost somewhere along the path since then. So when the waiter told me I had to wear a tie to be served, I tied my belt around my neck and insinuated I'd make a scene if he didn't take my order.

I'd paid the $3 cover charge because Jelly Belly,

the original Jazzman man himself, was appearing in person. I was going to see this legendary Beat musician and I wasn't about to be swept out like a mangy dog twice in one day.

At any rate, as I was listening to Jelly Belly, a Lady in Scarlet walked in. Her bright red jacket, puffy brown riding breeches, white gloves, knee high boots and black velvet helmet identified her at once as a high-brow fox hunter. She was the image of the lovely rich princess you would expect. Graceful, swan-neck, alabaster skin, perfect features, raven black hair, sexy violet eyes. She was stunningly beautiful..

Sour grapes fermented in my stomach and I could taste the bitter bile of unobtainable desire. She was with a person so bland that I don't remember if it was male or female. Languidly, the fox hunter gazed around, searching for an empty table. She actually shuddered like a horse when she realized the only space for them was with me. I grinned and pushed the chairs out for them with my foot. "Please, be my guests," I said.

"Not likely," she sniffed at me. "Let's go somewhere else," she told her forgettable companion. But the other person wanted to see Jelly Belly too and said they had perfectly good seats despite the present occupant, who after all was gracious enough to welcome them to his table. I grinned and grinned and said, "Sure, sit right down, folks." There were four chairs and my feet were in one and my posterior in another, leaving them no choice. They ignored me -- or I should say sneered at me with extreme prejudice -- obviously hoping I would take offense and leave. Not likely. I lit up a big black Cuban cigar and puffed smoke in the air. Why deny it? I was looking for trouble. The mere sight of this ostentatious fox hunter ignited a fire inside of me. You will understand why very soon.

I had just ridden 300 miles on a motorcycle and was exhausted. I let them know with a crocodile smile what they could do with their spurious, nouveau-rich, blue-blooded etiquette, as I perceived their indifference.

As she sat down, I noticed for the first time that she was wearing an aluminum back brace padded with white rubber. I assumed it was due to a spinal injury because her right eye kept twitching and her hands trembled violently. I said nothing, of course, for surely this was a tragedy of immense proportions to her.

Mostly I just wanted to be left in peace so I could sip my coffee, hate the snobbish society in Detroit, and listen to Jelly Belly belt out the blues. To my annoyance, however, she began talking. Though I didn't want to know, she informed her friend that she'd cracked a vertebra falling off her horse while chasing a fox. But she was going to take Daddy's advice and get right back on when her back was better, she said.

"I just love it! Tally-Ho!" she said to her companion; "The cry of the bugle; the baying of the hounds; the throbbing horseflesh between my thighs!"

Meanwhile Jelly Belly poured it out, "I loved ya, baby, I loved ya well! I loved ya baby, now I lives in Hell! Your skin was velvet, your eyes they shone, I loved ya baby, but your heart was a stone!"

But the scarlet lady wouldn't shut up. "When you see the red fox, your heart quivers like a

 rabbit in your breast! There's nothing like the thrill of the hunt!"

"Unless it's the thrill of the kill," I said before thinking. They stared at me with revulsion, disgust and hatred.

"For your information," she told me with an icy voice, "we don't kill the foxes. We keep

 them in' cages and let them go just before the hunt. They're too expensive to kill. The servants catch them when they go to ground. Now, if you don't mind, no one was talking to you, so please keep out of` our conversation."

"We lived in shanties, we lived with rats! " wailed Jelly Belly; "I bought you dem dresses, I bought you dose hats! You was my honey; you was my wife! But now you done stabbed me... and you's twisting de knife!" There were tears on his face.

The lady and Mr. Nobody kept talking, so I decided -- in the boldness of my youth -- to break up their little tea party. "So," I said to her, "you torture the same fox week after week? Maybe that broken back of yours is God's justice?" And I laughed loud like the hyena. The people around us grumbled and whispered. Perhaps I wouldn't be so blunt now, but that year -- as our Hippie prophets sang -- was the season of the witch and I was filled with an unspeakable anger.

Red in the face, she answered, "Again for your information, we have two foxes. And they live better in captivity than they would in the wild.

They live better than you. Now, if you don't mind, we'd like to hear the concert."

So that was the difference between fox hunters, I thought: one kind terrorizes the same fox over and over, while the gritty fox hunters I knew killed their quarry only once. I knew which was worse from experience,

"Well," I told her, "you kind of prove the saying that the higher you climb, the more you show your ass." That did it. The manager threw me out while the people in the pub cheered my ignoble exit. Later on, back on my bike, roaring toward Georgia, I remembered what it was like to be hunted and shot. I knew because of a game I used to play with my high school gang. It was called Fox and Hounds.

Whizzing down that endless nighttime highway . . . dreaming and a'dreaming . . . I remembered a particular day Pootaroot, Gooch, Willy, Roundtree, Marvin and I decided to play our Fox and Hounds game on a Saturday morning a Pigeon Swamp. The object of the game was for one boy to disappear into the swamp while the remaining boys hunted him down. And we took it very, seriously.

That Saturday morning -- in August -- we met at Pigeon Bridge. We brought football helmets, swimming gaggles, thick leather' jackets, denims and combat boots. We dressed like that because we' were armed and the shooting was to be real.

Pootaroot, Willy, Gooch, Marvin and I used Daisy BB rifles, which deliver a muzzle velocity of about 300 feet per second. It stung like a wasp to be hit, even through our armor.

The threat we most feared, however, was Roundtree's hand pumped Crossman pellet gun, which had a muzzle velocity of over 900 feet per second and could smash a Coke bottle at 100 yards. We all insisted he never pump it over four times and that he use BB's instead of pellets, like the rest of us. But Roundtree was treacherous in many ways. We knew he pumped the rifle eight or nine times when no one was loking. At that power, he could pierce our helmets -- he could burst an eye.

Roundtree, it's worth noting, was short and stout like his name. And he had a wide, lipless mouth that literally spread from ear to ear like a monkey when he laughed. Add to this a protruding jaw and tiny, piggish eyes with thick spectacles and you understand why so many of us called him a drooling ape behind his back.

His strange appearance was matched only by his strange behavior, which often was cruel and

morbidly funny. Fortunately, he was shorter and weaker than any of us -- so he was restrained in his bullying to kids in lower grades.

I used a Daisy Target Special, slightly more powerful than most of the other's lever action guns. I also cheated, for hidden beneath my jacket I always carried a C02 powered semi-automatic pistol made by Ruger. It looked exactly like a 9mm German Luger and had about the, same power as Roundtree's pellet rifle. I loved that pistol more than any other possession. Though it was chambered for .177 caliber pellets, I used BB's because the gun would fire fifty of them as fast as I could pull the trigger.

We got around our lack of hunting dogs by using the victim's own dog to track him down. The hunters held on to the fox's dog for thirty minutes then let him go find his master.

Because I had chosen the hunting ground, it fell on me to play the role of the first fox. I had to store my rifle in Gooch's car because -- as in a real hunt -- the victims unarmed and helpless. Of course.

I had chosen Pigeon Swamp for a number of reasons. First, I knew my way around because within it I often searched for snakes and other swamp creatures. Secondly, unlike the Okeefenokee, Pigeon Swamp was actually a marsh with muddy water no deeper than two feet and' literally filled with fallen dead or dying tree trunks which could be used as bridges. There were watersnakes and moccasins but no alligators.

Also on my side, I owned the dumbest dog in town and the only one afraid of water. Instead of looking for me, he might be thrown off track by a deer trail or muskrat lodge. Before I began the chase, Willy was holding Rex, who just sat and panted, staring stupidly around.

At the proper moment, I donned my black leather jacket, goggles and football helmet, ready to dash into the swamp. I had 30 minutes headstart before the others would come shooting for me. In typical adolescent cruelty, the fox wasn't considered dead until he shouted that he was giving up. Before that happened, our honor being at stake, the fox usually had to endure the stings of a dozen or more hits. But if hit in the face, the fox won by default.

Pootaroot made a sound like a bugle and I was off, jumping on the first long log and feeling my heart pound with excitement and an ele-ment of instinctive fear. I jumped from log to log as far as I could, then got into the water -- making sure not to drown my pistol -- and waded about thirty feet before jumping on another log and backtracking. I had made my way about a half mile into the swamp before I heard the second fake bugle call, signaling the beginning of the hunt. I could hear them shouting, "Kill the fox! Blow his head off!"

Trying to outwit them and my dopey dog, I decided to move sideways until I was out of the swamp and follow the creek down to the foot of Pine Mountain, from there to climb zigzagged, from tree to tree, in order to confuse Rex. It took me about twenty minutes to get out of the swamp and by then I could hear them closing in.

Using a trick I'd learned from raccoons, I climbed a tree and out onto a thick long branch as far as I could go before dropping back to the ground. Hopefully the dog would think I was still up the tree and the hunters would lose time searching it. Being August, the trees were thick with leaves. A bear could hide in one of them.

But they kept coming closer and closer. I knew Rex was leading them right to me despite every attempt I had made to cover my trail. When the hunters were close enough for me to hear them cocking their weapons, I chose the highest, thickest tree and climbed it to the top, hiding in the leaves. They found the tree within ten minutes and began firing indiscriminately into its foliage. I was hit on the jacket several times but they had no idea they'd hit me. Rex, my wonder dog, bawled like a real coonhound and stared right at me.

Following the dog's sight, they finally spotted me and began shooting fusillades of BB's into my hiding place. Several hit me on the hand and neck, which were quite painful, though I remained silent. Then I could hear them talking about chopping the tree down or setting it on fire. Willy actually began to climb up after me, but gave up when I began dropping pine cones on his face whenever he looked up.

Then I heard Roundtree say, "I'll get him down from there". And I heard him pumping that pellet gun ten times. He was an excellent shot and hit me on my left calf. It bled and I shouted down, "No fair! I'm bleeding!" They were supposed to immediately stop firing and shout up that I had won because Roundtree had actually wounded me. But I only heard them laughing, while Roundtree pumped up that awful air rifle again.

I knew I only one last chance, and that was to get down before he could pump up the rifle and run for a little cave I knew, carved in the bank of the creek. It was deep enough so you couldn't see the back, so I might be able to hold off long enough for them give up. I didn't make it and heard Roundtree's gun fire again, this time striking me on the helmet. Shouting, "He's shooting at my face!"

I dropped to the ground before he could pump up again and ran for the cave with all of them, Rex too, right behind me, peppering me with BB's. I turned my head around once and saw the hideous expression of their faces. This was no longer a game to them because, as I later found out, I had badly scraped Willy's face with pine cones. Foxes don't throw pine cones, so they had dropped the rules and decided all was fair.

I just made it to the little cavern and ducked inside. It ended about twenty feet away from the entrance and I was enveloped in darkness. I lie prone on the muddy bottom, facing the front, and took the safety off my pistol. Something had possessed these guys -- my friends -- and I was suddenly afraid of them. At least they didn't know I was armed.

Rex entered first, galloping up and slobbering on my face. He was a happy dog; he'd lead them right to me. He wasn't even wet, having circled around the swamp by listening to my footsteps on dry ground. It's difficult to describe my emotions at that moment. Except for Marvin, they entered the cave together, lined up -- Pootaroot, Willy, Gooch and Roundtree -- thinking as one to drag me out. Assuming I was unarmed, they fired one salvo after another, almost all of which bounced off my helmet or jacket. But it was Roundtree who fired the round that split my helmet. He must have pumped it twenty times to break through that fiberglass armor. I instantly realized his air rifle could kill me for real and I felt a horror I had never felt before.

I felt that I was no longer human but an animal at bay -- desperate and filled not only with fear but an outraged, irrational anger. Their silhouettes were outlined by the sunlight and the four of them came on relentlessly but unsuspectingly. Roundtree was the most dangerous and it was he who first felt the bite of my pistol as I pumped four BB's into his right kneecap.

I knew he was hurt because he grabbed his leg and fell, howling hideously. As the others were scrambling out, I shot each of them in the seat of the pants. "Game's over!" Pootaroot yelled, "Game's over!" No one argued and all was quiet. I gave Roundtree a dozen more brass balls in the rear as he crawled out on his belly. I thought it was over but they had one more trick up their sleeves.

They began lobbing into the cavern roll after roll of flaming camera film, which filled the hole with an overwhelmingly foul and unbreatheable smoke. To worsen things, they began throwing Cherry Bombs inside, whose explosions made my ears ring. Immediately after the onslaught, all of them except Marvin rushed inside and grabbed me by my arms, drug me out into the sunlight and threw me face down on the ground.

Game's over! It's a draw! " I shouted; "I'm out; don't shoot! And keep that animal off me!" I was talking about Roundtree, not Rex. Gooch stepped on one hand and Willy on the other. I was helpless. Someone grabbed my pistol. They lifted their boots off my hands but as I was trying to get up, someone kicked me over on my back and again my arms were pinned down. Willy stripped my jacket off, claiming he'd skinned the fox.

Roundtree was hopping around furiously without his rifle. Pootaroot had it and I felt its steel barrel pressed against my neck. Again I felt that inexpressible, mindless horror. I could feel my jugular vein pumping against the barrel. "I'm dead!" I cried out; "The fox is dead! I give up!"

But Pootaroot said, "You really are dead" and pulled the trigger. I panicked and jumped to my feet with adrenal strength that knocked all three of them down. I clutched my neck, feeling for blood. I had felt a gush of air and at that precise moment I thought he'd really killed me.

I've never forgotten that moment. Thinking how nice it must be to die and become a sunbeam was fine until my own life was in question. Then I wanted to live, no matter what! And I knew I would fight to the death before I allowed anyone to drive me underground again. I wish I could convey the anguished, intense terror I'd felt when the hunters were coming for me in the cavern. I wish I could reproduce inside of other people the madness and irrational, malignant killing instinct which arose in me. And most of all, I wish I could covey that heart stopping moment of terror when one knows death has at last grasped your throat.

To the others, I think, it was all part of the game. They laughed and joked and decided Roundtree should be tried for treason. He'd tried to grab the glory all for himself, breaking the code of the pack. I pretended to laugh too.

Roundtree was found guilty by our court, me being the key witness and my cracked helmet and leg wound prime evidence. We tied him to a tree, gave him a cigarette and a blindfold, then formally "executed" him with our firing squad of BB guns. Roundtree obligingly gasped and slumped against the ropes. So we cut him down and he sprang back to life and we all agreed it had been a magnificent hunt and we had all fought well -- even if a bit unfairly.

They seemed unchanged but I have never been the same since that day I was cornered. No, I wasn't hurt and this is not self-pity talking. Rather, I realized that if a man is an animal, then a non-human animal must have the same emotions. And it frightened me to have seen what joy some human beings take in inflicting fear and death. What is fun for the hunter is horror for the hunted.

I began to see my adolescent hunter friends less and spend more and more time to myself. When the time came for us to volunteer to fight in Vietnam, I joined the Navy while the others marched off to Southeast Asia as Soldiers. Only Gooch had the money to go to college. They probably dashed joyfully into the jungle expecting to track down, torment and destroy their quarry as they had the "fox" that day But I expect they forgot that the fox had been armed and that the Viet Cong were also armed. I believe this because Pootaroot, Willy and Marvin returned with terror etched forever on their faces. And Roundtree came back in a box.

I didn't change my mind about the necessity of death to evolve a higher consciousness, a finer being. I still understood that death was change, because time can't be stopped and if death were nothing, time would have to stop. I even retained the belief that my own death would be for the better once achieved and I would again become the sunlight from which I am made, ultimately to assume a new and higher sentient form.

But all that logic and philosophy had meant

nothing when I was lying in the mud with my pistol, waiting for the creatures determined to take my life. What is cosmology when you're up to your neck in reality? Whether true or not, my philosophy was meaningless when I myself had lost the veneer of civilization and knew intimately the animal fear and anger of the wounded beast at bay.

And I think that memory is why I insulted the beautiful lady fox hunter that brittle night so long, long ago.

Chapter Eleven
<u>The Death of the Viking Princess</u>

One afternoon later in 1970 I returned home from college and found a letter from Marvin in my mailbox. He had dropped out of graduate school, moved to Savannah, and was about to join the Navy. He wanted to "fulfill the obligation to my country in my own way," as he put it.

He believed a stint on board ship would be equivalent to Darwin's five years on the HMS BEAGLE, visiting foreign shores where he would observe the wildlife first-hand and develop certain theories about animal behavior he had formulated. Such a voyage would turn him into an internationally famous scientist.

He thought the Navy would recognize his genius and allow him to set up a laboratory on ship. "Well, at least a microscope and a table somewhere", he wrote. He added that he wasn't going in as an officer, but as a common seaman, as I had done. He didn't want to be burdened with the responsibilities of a Naval Commission.

He had talked things over with his recruiter and had been assured that any U. S. Navy ship's Captain would understand his noble interest in furthering the science of biology. "Don't worry one bit," the recruiter had told him; "The Navy is for individuals like you. You'll have plenty of freedom and time to do what you want. The Army wouldn't even let you own a microscope."

The outrageous promises of the recruiter didn't surprise me, but I was somewhat amazed that our brilliant Marvin would fall for them. I thought I knew what lay in store if he carried out his plan.

My recruiter had promised that I would be automatically selected for training as a Navy journalist. There was no doubt about it, he had said, a young man with my talents would be recognized, and I would spend my tour of duty in South Viet Nam writing U. S. Navy press releases. In other words, I would become a military combat journalist. The excitement of going to war thrilled my naïve and romantic teenage soul. I would write for newspapers and magazines on my own time, become a famous war correspondent.

Naturally, after my hitch was over, I'd be quickly snapped up by a major newspaper or national magazine to cover other wars for them. I'd write about the enlisted men and the harrowing experiences they endured day to day. I'd make grown men weep with tales of sailors and Marines maimed or killed by the monstrously cruel enemy. The Navy would begin my career as a war correspondent with a bang! I signed on the dotted line and was shipped to boot camp.

I was a small-town boy and believed every word, as Marvin had. But the recruiter's promises meant nothing once I was in. Boot camp made it very clear that I was a loathsome, dispensable grub whose only purpose was to follow every command immediately and without question. If my Captain ordered me to block an incoming missile with my body, I would do it. If he ordered me to scrub the deck with a toothbrush, I'd do it. I graduated a very bitter young man, filled with a determination to live alone and only for myself, an individual.

I wanted to talk to Marvin before he signed anything. As romantic as it sounded, serving four years on a Navy warship during wartime was actually a grueling, bitter lesson about the conflict between individualism and being an anonymous component of a team.

The Navy was oblivious to my dreams. Nor did they give a hoot about what I'd been promised. Instead, they tested me and decided I would be best suited to serving the Navy as a Yeoman because I could type. Yeoman is the naval term for secretary. Right after boot camp, I was assigned to the clerical office of the USS RALEIGH LPD 1, an amphibious attack landing ship, where I was to type reports and business letters for four months.

Once on board, I never left her until my hitch was over. The USS RALEIGH was, in my eyes, a majestic fighting vessel. She stretched over 400 feet bow to stern and 50 feet from port to starboard. She bristled with eight "three-inch, fifty-caliber pom-pom guns," designed to sink other ships, shell enemy land positions, or blow aircraft out of the sky.

The front part of the ship resembled a large destroyer, but there the resemblance ended. Her stern was a landing deck on which sat four black, sinister-looking Cobra combat helicopters. And inside her huge hollow belly was a well deck containing eight armored landing craft, each mounting 50 caliber machine guns.

During a battle or maneuvers, the RALEIGH would fill her ballast tanks and sink halfway into the sea. When the well deck was full of water, the hinged, flat tail-end of the ship could be lowered into the ocean like a ramp, leaving a flooded exit. Through this opening the landing craft would swarm out into open water, each filled with dozens of fighting Marines ready to invade.

The ship boasted a crew of 400 sailors and officers, plus 1,200 Marines and their officers. So 1,600 of us lived and worked together every day, all crammed at night into tiny compartments packed with collapsible canvas bunk beds and Lilliputian lockers for our clothes.

For months I rebelled against being assimilated into the crew, feeling that I was nothing more than a worker ant in a giant metal nest brimming over with other worker ants. I got in much trouble, especially when I jumped ship in Jamaica and tried to disappear, to become (in my imagination) a beachcombing writer.

But when I returned two days later with my tail between my legs to face Captain's Mast – the Navy equivalent of misdemeanor's court – the Commander of the ship took me aside and said I was not to be punished because I was only 17 years old and had not yet learned the importance of team work to accomplish great tasks like winning a war.

I repeatedly put in requests to attend a

Navy Journalist's school, but they were always denied. The personnel officer would sneer and snicker at me, saying every time, "Welcome to the USS-R! There ain't no virgins on this boat because you got screwed when you came on board!" He thought this very funny and would howl with laughter at his own joke.

On the RALEIGH, we "salts" were all individuals, of course, but together we formed one massive, interdependent crew. And no one sailor or Marine was allowed in any way to be different from his fellow crewmen. Our uniforms of the day were dictated by the ship's Captain, down to the last detail, including our skivvies and socks. Every minute of our time was preplanned. We knew exactly what we would be doing at any given moment, even brushing our teeth or taking our showers.

The only deviations from the "Order of the Day" were the Battle Station drills, which came unexpectedly two or three times every 24 hours, often during the dead of night. This involved all 1,600 of us donning combat gear, helmets, and life jackets, then scrambling over the decks like ants for our stations, trying to avoid mass confusion and be ready for combat within three minutes. If we failed, the Captain would repeat the drill over and over until we met the specified time limit. They often went on all night.

I was stationed at a pom-pom gun at first, where I handed the heavy shells to a Gunner's Mate. Later I was transferred to the bridge when the ship's photographer was caught drunk on duty and sent shamefaced to the deck crew as a "deck ape". The moment I heard this good news, I requested an audience with the Captain and was subsequently appointed the official ship's photographer.

From then on, I spent most of my sea tour with the Flag Officers and Captain on the bridge. Jumping the chain of command was risky business, and I might easily have been disciplined instead of rewarded for going directly to the Captain. But striking at the right moment paid off. After that, the Navy was as exciting and thrilling as the recruiter had promised.

Very slowly, I began to realize the selfishness of my determination to be an individual on a ship of war. Why should the Navy concern itself with making me a successful war correspondent? What good would it do the USS RALEIGH in carrying out her mission?

Two things happened which changed my viewpoint from "I" to "us". First, a Norwegian luxury liner named the VIKING PRINCESS caught fire in the Caribbean Ocean and had to be abandoned. Though only one passenger was killed, the ship herself seemed to all of us a living creature which had to be saved. The RALEIGH sped to her rescue.

It was a terrifying sight as we drew near that beautiful ship ablaze. Explosions inside of her would send massive pillars of fire and black smoke into the sky. All the paint had been burned off her, leaving only a blackened metal hull. The heat was so fierce, the RALEIGH could get no nearer than 100 yards or so. Those who had volunteered to board the PRINCESS to fight the fires would have to approach in landing craft and scale her hot shell with ropes and grappling hooks.

I was to go with the fire fighters, documenting the adventure with my cameras. But my Captain said to me the mission was not an order, and I could remain safe on the RALEIGH if I preferred. For a minute I was painfully unsure, but ultimately chose to go. I would do my bit in the saving of the VIKING PRINCESS.

The seas were rough, and we had to climb down nets thrown over the side, jumping into the landing craft just after the swells peaked and the craft was falling back down. Timing was vital, for if we jumped when the landing craft was being lifted by a wave, we could break our legs. I was frightened before I'd even come near the burning ship. But I jumped and landed unhurt.

At the PRINCESS, enveloped already in an overwhelming heat, the coxswains threw their grappling hooks, and we climbed aboard, the hot ship's hull burning the soles of our boots. I began clicking pictures of the fire fighters as they scrambled around, setting up hoses with ten-foot metal nozzles, each spewing sea water from pumps on the landing craft. The noise was deafening, explosion after explosion.

But I wasn't prepared for what happened next, never having been on a burning ship at sea. As the sea water cooled the decks and bulkheads, the metal would contract with tremendous booms, hurting our eardrums. But then they would expand again, booming again, and throwing men and equipment everywhere. Five fire fighters were thrown overboard; another broke a leg when a loose hose snaking over the deck smashed against his knees.

It was then that the most peculiar feeling came over me. The crewmen were fighting a monster fire devouring the ship, while I was standing aside snapping my camera. I felt at that moment that I should be one of them instead of an observer. So, slinging my cameras behind my back, I grabbed a hose with four other men and began fighting, struggling – one of the team – to beat this thing trying so hard to destroy us, throw us overboard, break our bones, or burn us into charcoal.

Four hours we fought the fire before it finally began to recede and hide deep in the bowels of the PRINCESS. We followed it down with our hoses and axes, forcing it further and further inside until we had it cornered and almost dead in the engine room where it had begun. Once confined, the B'swain's Mate ordered us to retreat so he could flood the compartment. Water was rushing in as we climbed the ladder out, slamming the watertight port behind us and sealing it shut. We'd won.

I was trembling and shaking, but so were the others as we again reached the fresh air of the main deck. I resumed taking photos until I saw the only fatality lying on his back, wearing only a swimming suit, by the liner's pool. Over half of his body was charred black; the other half was roasted and smelled so much like broiled beef many of us vomited.

I put my cameras away again and joined the Hospital Corpsmen, helping them lift the body

– whose arm broke off – into a body bag. I noticed that several of his fingers and toes remained by the pool. We had to gather up all the lost parts and stick them into the body bag before zipping it up. Then I helped them carry it to the side, where we had to lower it with ropes into a landing craft.

For the first time in my life (and, sadly, one of the last), I felt a part of something, a crewman among fellow crewmen. And I felt proud that together we had conquered the fire and saved the PRINCESS from sinking. My individuality had been absorbed for a short time, and I felt very, very good.

The second incident came when the USS RALEIGH was ordered to the island of the Dominican Republic to help several thousand U. S. Marines invade the tiny country. Though I didn't know it then, our mission was to rid the Republic of the communists about to take over the government.

I was at my Battle Station on the bridge as we, along with dozens of other ships, dropped anchor about a mile off shore. As I stood there, I could feel the RALEIGH filling her ballast tanks and sinking lower into the water. I could feel the giant stern door opening and lowering into the sea. I could hear our Marine officers shouting orders as their men filled the landing craft and began to leave the ship to circle in figure eights while all the others were launched.

The pom-pom guns fired in a constant roar of thunder, shaking the whole ship. I could see the shells exploding on the coast, blasting palm trees into smithereens. Only scared trunks remained, at the top of which a few very dead snipers hung from ropes where they had tied themselves in the branches.

Then the Marines hit the beaches, wave after wave, thousands of them running across the sand. A few fell, only to jump back up and continue the sprint for cover in the jungle. Rifle and machine gun fire was everywhere. The small resistance force melted before the Marines, and we watched as they either ran for their lives or stood with their hands in the air, grinning like baboons.

I was snapping photos through a telephoto lens. But again I had that wonderful feeling of not being just myself, but part of an invincible, powerful force in the midst of battle. I realized in the fighting that here was something finer, more noble, more real, more natural than the romantic notions I had nurtured of being a loner, a stranger to everyone, an observer and recorder of the heroism of other men. I wanted that fine feeling to last forever.

Unfortunately, I never found it again since leaving the Navy.

For me, living on a warship was much like being married to a demanding woman. You think you hate her until she is no longer part of your life. Then you mourn your loss and appreciate too late the lady you've lost. Decades later you find yourself dreaming of being back with her, only to awaken with a sadness wrapped around your heart like a strangling vine. So it is with the USS RALEIGH and me.

Chapter Twelve
The Courage of Field Mice

The memories of the SS VIKING PRINCESS had been sparked by Marvin's letter. I had no doubt that the Navy experience was good for normal men. But not for Marvin. I felt obliged to warn him that a sailor either gives his soul to the ship or he is systematically driven insane.

Marvin was vain and individualistic to the extreme, displaying idiosyncrasies which would certainly lead to disaster. His resentment of authority was almost pathological, as I had seen that night when he shoved Pootaroot, the self-appointed leader of our high school pack, face down in the dirt. He was a rebel in every way, demanding everyone respect him while showing very little respect for others. He simply was not Navy material, in my seasoned opinion.

I thought his only chance was to get back in graduate school before he joined the Navy or was drafted. So I called him in Savannah and tried to tell him what he was asking for.

He was only one semester from earning his Master of Science degree in zoology. He'd been the only one of the pack at Meriwether High to win a full scholarship, and he was about to throw it away. Sure, the G. I. Bill would help him finish his education when he was discharged. But there was a very real danger that he would be so changed by the Navy that he might become a totally different person, losing his one chance at a career in science.

I wanted to tell him that the military experience is an anvil on which the metals of our youth are hammered into shapes we could never have imagined.

But Marvin was stubborn and insisted the Navy would test him and discover his great intelligence and value. They would surely provide him with a ship's laboratory because they'd realize mankind's fate depended on scientific men like himself. I laughed aloud at this, making him only more determined to prove me wrong.

Despite my advice, he joined anyway, and I wasn't to see or hear from him for another four and one-half years. But eventually he returned, finding me ready to say I'd told him so. I wasn't to have that pleasure.

Marvin came to see me shortly after his return. He came over beaming, suntanned, and carrying himself with a haughty bearing I had never seen in him before. As I sat stunned and astonished, he told me about his four years.

Laughing, he told me how he had avoided the hardships of both boot camp and Officers Training School by volunteering to be the company clerk. While the others were learning to be sailors or officers, he merely kept records in an office. Most of the time he had only to sit back, his feet on his desk, reading books and watching the clock.

After receiving his commission as an Ensign, he was assigned to the aircraft carrier USS

Enterprise as a science officer. There he shared a laboratory with two other science officers studying ocean conditions and marine life. Their job was to determine how such knowledge might benefit Navy underwater demolition teams and pilots stranded in the sea.

Yes, they had given him a microscope. He had finished his remaining semester of graduate school at the University of Maryland while he was supposed to be working in the lab. For three years, six months he had been doing precisely what he wanted and what his recruiter had promised: visiting exotic foreign shores around the world and studying each land's topography, ocean conditions, and wildlife.

Upon achieving the rank of Lieutenant as his tour was about to end, he had contacted his former graduate school and discovered that the experience he had gained would be credited to his academic work for a Ph.D. when he returned. As he was talking to me, he revealed that he had already been reaccepted and should have his doctorate in marine biology within a year. No fool, he had written his dissertation on the migration habits of bottlenosed dolphins – while he was supposed to be searching for survival techniques for Navy pilots stranded in the ocean after being shot down.

With a big smile, he also told me he had a curator's job at an oceanography institute run by a Florida university waiting for him when he'd received his doctorate degree.

I had little to say, so congratulated him and let him talk on. I didn't hear much else because I was lost in thoughts of my own. He had avoided every hardship. He had never fought, never considered – even for a moment – himself part of the crew. He held his ship and its crew in contempt. He never knew what it was like to love one's ship. Despite his proud talk of how he had used the Navy to his own advantage, I felt sorry for him. He had outwitted himself.

Marvin had never been forced to make a frightening decision. He had never known what it was like to stop feeling alone in the belly of a great ship, as if swallowed by a whale. He had never felt that wild, gleeful joy of choosing courage over cowardice, fighting a dangerous enemy and winning.

After he left, still smugly insinuating how clever he had been, I got on my motorcycle and rode through the night into the deep woods. At some small hour of the morning, I don't know which, I left the paved road and followed a dirt trail to a pond hidden in the trees.

As always, the forest was cool and calm. I found a little clearing, where I made a campfire and sat beside it to think about this question of courage and cleverness. Leaning back against a pine tree, I watched for meteors in the sky. Nine times the shooting stars traced across the sky, and nine times were swallowed by darkness. The campfire too threatened to be swallowed up, but remained alive so long as I fed it the wood it demanded.

I remembered the Goat Man from so long ago, who had roamed the pine woods with his wagon and goats, avoiding civilization as if it were in the grip of a terrible epidemic. I no longer wished I could ask him if I could travel with his goats, carefully living a life unencumbered with hard choices or struggle for existence.

Though I was to change several times back and forth during the years that followed, at that moment I didn't want to be a goat in the forest. I was to become one through circumstances later on, but that particular night – remembering the odd warmth of losing oneself in a fighting team – I wanted to be again a sailor at war on the USS RALEIGH.

Solitude was blissful that night, and soon I was feeling a freedom which made me forget – for a while – the mocking laugh of Marvin as he described his premeditated dereliction of duty. The wind rustled the leaves and cooled me. The ten thousand insects and tree frogs continued their serenade, calling lovers to meet them in the night.

I was before the Navy a goat in the forest and wanted to remain forever cloaked by protective branches. Ultimately I would become one again because I couldn't find human comradeship and a common enemy in civilian life. In fact, I was to find civilian life in America as brutal and lonely as others find only in prison. But that night I was disturbed by the idea Marvin had expressed that it was foolish to take chances, to make dangerous decisions, to face uncertain tasks, to band with other people, to fight enemies of a country which really had no substance . . . was, in fact, a mere collective dream of millions of individuals.

I wondered if Marvin was right and I had been an idiot to fight the fire on the PRINCESS and give myself to the battle for the Dominican Republic. If I was so wrong in living dangerously, why did I feel so good when doing it? Why had the sailors and Marines I thought I detested suddenly become my closets buddies when the common enemy was engaged?

I began to doze on and off. The fire died slowly with complaining crackles and spitting noises. Eventually all was dark except for the starlight and a half moon. The symphony in the forest went on, however, like a lovely lullaby remembered from those days in a crib when reality was milk and a mother's face.

I was startled into consciousness when I felt something crawling up my arm and sniffing my neck. It was a field mouse, I knew, and didn't move. I let him crawl into my hair, where he snuffled about a while before taking alarm and suddenly jumping of as he realized I was awake.

He froze when he hit the ground, staring at me. That little field mouse, so unlike the mice found in human homes, had eyes so big I could see them in the dim light gleaming like black pearls. And in those eyes I saw something, like a fortuneteller gazing into a crystal ball. I saw the answer to Marvin's troubling insinuations that avoiding dangerous decisions and fights and comradeship was the wise man's way of survival and success.

The animal's eyes mesmerized me. It was nearing dawn, and I began to think in terms of light instead of dark. No longer was he "a poor little mouse", but a being, a living creature, warm and content, unknowing of his certainly violent end. But what did that final instant of being mean to a field mouse? For him, there was only the moment, and it was generous enough. I saw for the first time the intense vigor and vitality dwelling within him.

He reminded me of two other field mice I had watched years before. It was dawn, and I was at another camp, alone and wondering about the incomprehensible Nature of life. Those field mice had also come close to my camp and were warming themselves by my dying fire. But earlier that night I had captured a Copperhead and had allowed it to move around my camp, so I could watch the serpent's movements.

The viper had coiled near the fire and had remained motionless and almost invisible for over an hour. And as I watched in the morning light, one of the field mice began to investigate around the coals. He came near the viper several times without mishap. But eventually he touched the silent snake, who struck instantly but missed.

The frightened field mouse jumped straight up into the air with a squeal. But when he touched the earth again, he didn't run away as I expected. Instead, he attacked the viper's head, delivering a vicious bite into the snake's skull.

The Copperhead was unhurt but obviously shocked by the mouse's hot temper. As the snake was crawling away, the other field mouse joined his partner, and together they launched an attack on the viper's tail. The two of them actually bit off about an inch of it before scampering back to the fire. Trembling with anger and excitement, the field mice sat on their haunches and watched as their deadliest enemy retreated.

Who would have expected such courage in field mice? The memory of the heroic battle filled me with a strange wonder. It was incredible to me that the dragon didn't always win. Sometimes the mice won through sheer bravery and the boldness beating in their vulnerable breasts. Having seen death in the viper's eye, they had decided somehow in their instinctive minds to die fighting. Rather than run, they had struck a blow, and that had given them their victory over the venomous fangs of the snake.

Had they been clever, at least one would have surely died from a second strike. Only the dangerous decision to fight saved him.

This I remembered as I stared into those black pearl eyes of the field mouse that night Marvin had ridiculed facing perilous choices and deliberately engaging menacing enemies.

But as the sun rose in the east, casting a red glow over the forest and reflecting off the pond in a rainbow of fiery colors, I felt the courage of the field mouse in my own marrow.

One had to fight, make decisions, and struggle, always keeping a splinter of hope ablaze. And when the vipers come, though they may have slain a million like us, nevertheless strike back at our horrifying fate. In that process, some of us inevitably must emerge victorious and free. There is always that chance. Those of us who survive reach a plateau unexpected by those who run away. We are stronger and wiser for the courage we have generated in ourselves.

As for Marvin, his cleverness eventually failed him, despite his foresight and energetic ingenuity at avoiding trouble. He did indeed get his doctorate and the curator's position he'd been promised. But a scandal later broke out at the marine institute when the severed head of an endangered sea turtle species was discovered in a trash can. The resulting media publicity forced him to resign and accept a job teaching math to seventh graders at a public school.

Teaching is an admirable profession, but I doubt if Marvin appreciated his new position. Perhaps he felt slighted when the school administration refused his request for a microscope, for I heard later that he had quit. I don't know what he's doing now, but fortune being as unpredictable as it is, perhaps Marvin has again risen to the heights in the scientific world he had known before. All he needed was the courage of a field mouse to fight back when providence threatened to destroy him.

We are bewildered beings on this earth. We don't know when to be goats in the forest or when to be mice facing vipers. Each of us must choose again and again, in circumstance after circumstance, each one different than the one before. Dangerous or not, choices have to be made. That is the albatross hanging around our human necks.

Perhaps it's in this very act of choosing that our greatest hopes lie. Maybe the key to finding an equilibrium in this roller coaster world is always making those decisions which most empower and most embolden our otherwise cowardly hearts.

Chapter Thirteen
<u>Wisdom in a Serpent's **Eye**</u>

It is cruel and certainly ironic that in our ignorant youths we are forced to make the decisions which will make or break our lives. As adults, we look back astonished that we could be so foolish and shortsighted. But life what it is, there is no spinning back of the clock. We are doomed to live out the shattered illusions of our imperfect adolescent vision.

This sad observation was on my mind yesterday as I followed a creek through the summer woods, looking for a shady private place to lay down and relish the aloneness I have come to love more than any woman, any ambition, any memory, any battered dream.

My companions in the forest these days are cameras and books instead of human beings. I was like this as a young child, and I am like this again as I approach the waning of my years. But, like most people, there was a brief flash of unwise hope in my young manhood which envisioned a life of romance and adventure. As with so many others, pursuing this mirage blinded me to the facts of existence I was later to learn at such great cost.

I am not the exception but the proof of the rule. When I was 23, I dropped out of college after my second year and moved to Atlanta. My plan was to follow my friends, who had all transferred to

Georgia State University in order to join the general extended party of cheerful beer-drinking bouts, abundant sexual encounters, and plentitude of beautiful girls with whom to fall in love.

    I moved but found none of those things. My friends were there, alright, and when I saw them I heard fantastic stories about the so-called sexual revolution. Willy told me he was living with a girl, as did Pootaroot and Gooch. They told me about outrageous parties where semi-nude college co-eds painted themselves in fluorescent dragons and danced underneath black lights, glowing and shimmering, seductive and alluring . . . and best of all, available.

    It was the Season of the Witch, those hippie years, and life was said to be one long magic celebration. At least, that's what was shown on television and claimed by my friends. The illusion was only strengthened by the books I was reading then about the tragic lives of artists and scientists who lived in hovels, ate nothing but day-old bread, swirled in milieus of ecstatic hungry women, discussed philosophy with genius friends, and suffered to express the brilliant visions illuminating their minds.

    But all these things somehow eluded me. Oh, I found the hovels, ate the black bread, and suffered (mostly from boredom). But for the life of me, I just couldn't seem to find those liberated, sexy, available girls. I couldn't find that loving live-in lover. Instead of philosophy, the people I met talked about football and cars. And though I did attend a few wild parties, I only sat miserable and alone while others danced their drunken tarantulas.

There was also the ugly problem of making a living. It takes a lot of money just to be poor in a giant American city. To live without any money at all usually results in nights spent on mean streets with winos and armed teenage hoodlums.

After a few months of sporadic manual labor, I accidentally landed a job as a keeper at the Atlanta Zoo's Reptile House. At the time, I could hardly have realized that this stroke of luck was to be my one and only chance of becoming the authoritative naturalist I had always expected to be.

I had been obsessed with the primeval savagery and horrible beauty of reptiles since the age of five. Until I went to college, there was never a doubt in my mind but that I would one day be a herpetologist, a professional specialist in reptiles and amphibians.

How I wish I had known then that I was already a professional herpetologist. With just a little patience and a little night school, I could have risen in my field at a slow but sure and secure pace. But I was 21 and couldn't wait.

Though I didn't know it at the time, the Reptile House was my new ship. I loved it in the literal sense, just as I had loved the USS RALEIGH three years before. I admired the Curator, my new Captain. And I respected my fellow keepers, for together we were the crew of this magnificent ship dedicated to the care, display, and preservation of reptiles.

But being 21 is an age of horrible irony. We rarely recognize good fortune and instead curse her for getting in the way of our dreams. Now that it's too late, I know that if I had remained where I was, today I would be a Curator of Reptiles instead of a beaten, jaded scribbler.

The Reptile House was divided into six departments: venomous New World snakes, venomous Old World snakes; harmless snakes and lizards; turtles; crocodilians; and amphibians. I was assigned to the wing reserved for North, Central, and South American pit-vipers.

It was heaven for me, and I didn't recognize it. I was proud not only of my charges, but of my nattily pressed khaki uniform with its epaulets, patches, military pockets, rakish cap, and knee-high black boots. I also proudly carried my custom-made, patent leather handled snake hook, swaggering like a German Field Marshall carrying a commander's baton.

Under my care were over 25 species and subspecies of rattlesnakes, ranging from tiny 12-inch pygmy rattlers to enormous eight-foot-long eastern diamondbacks. I had sidewinders and timbers and prairie rattlers. I had three big water moccasins and four copperheads.

Also in my keep were Central and South American pit-vipers, strange slender creatures with rough, file-like scales and evil slit eyes. The pride of my section was a nine-foot long Panamanian bushmaster, the second largest venomous snake in the world, sporting two-inch long fangs and the vicious disposition of a caged panther.

Due to space problems, I also had in my charge a thirteen-foot-long King Cobra from Southeast Asia. He was an impressive creature, with a body as thick as my arm and a head as large as my hand. In most people's eyes, I'm sure he was considered a monster. He could deliver through his half-inch fangs enough paralyzing venom to kill any animal walking on this planet. There are documented accounts of King Cobras killing adult elephants by striking them on the end of their trunks.

Only one human being is known to have survived the bite of King Cobra, and that is a professional snake handler named William Haast in Miami who had for years before been building an immunity by injecting himself weekly with diluted cobra venom. When he was bitten, however, he came within inches of death, despite the tolerance he had developed.

But it wasn't the size or lethal bite which enthralled me. It was the snake's uncharacteristic, uncanny intelligence. Whenever I watched him through the window of his cage, he would rise three or four inches in the air and stare into my face with bright, glittering black eyes. He seemed to be as curious as I was.

Most herpetologists agree that the King Cobra is the most intelligent of all serpents. My charge was no exception, for he knew the difference between the Curator, the other keepers, and me. He knew which days his cage was to be cleaned and which days he was to be fed. And he knew that he was a king, for he never once showed the slightest fear of us or any other animals which came under his fearful scrutiny.

I developed a peculiar relationship with the King Cobra and would open the back of his cage every morning to pet him with a long snake hook wrapped in rags. He enjoyed these rub downs and would shiver with delight, sometimes even turning upside down to have his belly scutes stroked. He was so tame, in fact, that often I was tempted to caress him with my hands behind the large venom sacks on his head.

The Curator strictly prohibited the handling of cobras by keepers for any reason, apparently due to their clever deceptions and extraordinary killing capabilities. That was his responsibility, and he enforced the rule with an unforgiving policy of firing anyone who was caught. But I was proud of my knowledge of reptiles and the experience I had gained through years of raising them as pets. So I broke the rule everyday. I just couldn't resist the thrill of petting a beautiful, majestic killer.

Doing stupid things came to me naturally in those green years. I was also very good at misunderstanding the motives of other people. As time passed, I began to resent the Curator because he never seemed to notice what a knowledgeable herpetologist I was. When talking to him, I would always use the scientific names of the reptiles. I would make suggestions about the proper way to handle problems, like a snake who refused to eat or a crocodile who was becoming too aggressive. But he always ignored me.

It never once occurred to me that the Curator must have respected my knowledge and abilities, or he would never have put me in charge of a venomous snake section. It never occurred to me that I was dealing with a quiet, dedicated scientist who avoided unnecessary talk and was wary of being too friendly with his subordinates. Nor did it occur to me how pretentious I must have seemed to him.

Instead – damn me – I got it into my head that the Curator was jealously guarding his high-level position. To me, he was throwing it into my face that I lacked a degree in zoology. Tings got worse as I came to feel more and more unappreciated, more and more separated from the respect I thought I deserved. I felt he and I should be equals and the best of friends, bound by our mutual fascination of reptiles.

As I worked with my beloved reptiles everyday, I began to daydream about returning to college, getting a master's degree in zoology, and retuning to the Reptile House with both academic credentials and experience. Then, I thought, the Curator would respect me and I would finally be a professional herpetologist.

These plans were not unreasonable, and I might have accomplished them with ease. But I was too impatient to follow through. I wanted it all, and I wanted it now.

The only appropriate word to describe the King Cobra is "majestic", for he is without doubt the most regal creature to ever reign on this astonished planet. He is not only royal, but fearless and always ready to battle for his existence, his freedom, his integrity, his right to be King.

So it was painful for me to see the King

confined to his cell like a mere snake. He was so much more. His spread hood was a flag, on the back of which was emblazoned his coat-of-arms. His body was covered with bronzed shields, each glistening with a thousand prismatic hues. His eyes were glittering jewels, black as onyx, noble and sublime. I could stare into them for hours, mesmerized and charmed.

He was so long that he could lift his head six feet into the air and stare a human being right in the eye, unblinking and assuredly, without bluff. There was no other serpent to match his magnificence in the Reptile House. The boas and pythons were larger, but less graceful and lithe. There was nothing ungainly about the King Cobra; his elegance was as instinctive and quick as the wind.

As I grew more and more entranced with the giant cobra, I also became more and more reckless. Eventually, when the Curator was out of the building, I would open the cage and stroke his muscular, sinuous coils with my hands. I never got close enough to the deadly head to give him a chance to strike, should he suddenly decide to do so. I was careful, but at the same time very young and very foolish. I would not risk my life so quickly now.

But though I came to love and venerate the King, there came that time every evening when I was forced to lock up and go home. I dreaded those times, for a different reality took hold of me, and I despised the lonesome, often loathsome mean streets of Atlanta.

After work, I would return to my hovel and

brood until I could stand it no more and walk to a students' pub called The Stein Club. There I would hob-nob with other young people with dreams, but in their case dreams of becoming famous rock and roll stars, artists, actors, poets, novelists, dancers, and every other glamorous, if unrealistic, profession imaginable.

Then I would forget all about my daytime pans and fall under the spell of seeking quicker fame and glory. Though I haven't mentioned it before (out of embarrassment), I too had an impossible dream. After a few beers, this dream no longer seemed impossible, and I found myself telling the other dreamers that I was going to quit the zoo and carve out my own destiny.

I'd might as well admit the mirage shimmering in my mind. I wanted to be an adventurer, leading expeditions into dangerous jungles in search of exotic reptiles. In this dream, I would make my living by selling the snakes I captured alive to zoos. And I would become famous by writing about and photographing these great adventures, producing accounts which I would sell to magazines or compile into best-selling books.

In the house I would build, there would be one room reserved for a King Cobra. I would watch him, fascinated and exhilarated to have such a splendid roommate.

During these beer-soaked nights, I would beat myself for being too cowardly to pursue the dream. I'd make plans to save enough to go to the Southeast Asian jungles, where I would live with a tribe of natives while studying the King

Cobra in the wild. Certainly "The National Geographic" magazine would snap up my story and photos. The princely sum I'd earn would finance yet another expedition. Nothing could stop me! I'd open my own reptile institute!

How dull and commonplace my keeper's job seemed then. I felt I was an insignificant subordinate and nothing more. The deeper the night, the more firm became my resolution to walk off my job and leap into the unknown jungles of the future, where awaited my fortune. Like the cowardly lion, all I lacked was nerve.

Now that I am past the age of seriously cultivating a career, I know that it was this adolescent pride and romanticism which led me to catastrophe.

Trouble was inevitable. Neither conflicting dreams nor beer is conducive to clear thinking. So it's no wonder that I decided once and for all to force matters to a head.

Instead of quitting outright, I decided to prove to the Curator that I was as much of a herpetologist as he was. By my fuzzy logic, once he admitted this to himself, he would not only treat me as an equal, but get to like me and perhaps even promote me to Assistant Curator. Then, when he was promoted to Director, I would assume my rightful place as Curator of Reptiles at this internationally respected zoological park.

And if it didn't work, I'd have no choice left but to seek out my secret dream of pursuing the elusive cobras while exploring the Asian jungles. I would leave it all up to Fate, that fickle woman, and take the path she indicated. Either way, I thought I would win.

It all seems so childish now. Despite the rule against handling cobras, I decided I would prove my expertise in herpetology by taking the King from his cage and demonstrating how professionally I could control him. I would easily keep the creature at bay with my snakehook, I thought. And as I manipulated the most dangerous snake in the world, I would meanwhile be expounding my tremendous knowledge of the cobra family.

But the best-laid plans of man and snake often go awry.

I had manipulated vipers and pit-vipers hundreds of times and felt no fear of controlling a cobra, even one of such great size. The Curator made his rounds everyday like clockwork, visiting my section at exactly 10:30 in the mornings. So at 10:20, I opened the King's cage and gently lifted about five feet of his head and body out with my snakehook.

He balanced himself carefully, at first gazing curiously into my eyes, then at the ground, then into my eyes again. He seemed to be thinking things over. I was careful to keep his head far enough away from me to prevent a strike. I thought I had him under control, and the Curator couldn't really accuse me of handling him because my hands would never touch the snake.

But when the Curator walked in and saw me with the King Cobra, he froze in his tracks. "Put him back!" he whispered in a coarse, frightened voice; "This ain't no circus!"

As the King Cobra's head and about four feet of his body swung in mid-air, the rest of his body glided out of the cage before I could react and coiled on the ground. His great head rose to the level of our eyes, and his hood spread in warning. Every breath he exhaled was a loud, long, ominous hiss.

I was suddenly paralyzed by a stark terror I had never expected. The King looked first at the Curator's eyes, then back at mine, slowly and with what I swear was a condescending amusement.

Then for several minutes his eyes fixed only on mine, and something primeval and archetypal passed between us. Man and cobra had been through this place before, and an instinctual imperative to scream grasped my throat. But I couldn't speak or move. I thought he would certainly strike at me or the Curator if I so much as breathed.

It was then I discovered of what metal the Curator was made. Swiftly and seemingly without fear, he used his own snakehook to lift the King's head – every so gently – up to the door of his cage. The mighty King glided back into his quarters and climbed a tree we had in there for exercise. The Curator had the door shut and locked before I could release my breath.

The Curator was justifiably angry but stood silently looking at me for a long time before speaking. Finally he said, "That's the dumbest thing I've ever seen anybody do in my life. I don't care so much that you almost got yourself killed, but that cobra's worth $12,000! Don't both to come in tomorrow. You're finished at the Atlanta Zoo."

All at once, the enormity and sheer stupidity of what I'd done hit me. In less than a millisecond, I became a mature man with mature thoughts. But it was one millisecond too late for apologies or explanations. All I managed to say was, "Couldn't I work in the frog room until you trust me again? I was just trying to earn your respect. I wanted to show you I know my snakes."

He cooled off a bit when he heard this. He almost seemed sorry for me. "You were showing promise, Richard," he told me; "I thought you had what it takes. But handling snakes is too dangerous. We're scientists here; there's no place for heroes. You broke the one rule which can't be overlooked, not even once. So get your stuff and leave the uniform in your locker," he said, turning on his heels and walking out. "And leave that snake alone!" he shouted as the door was closing . . . closing forever on my future dreams to be a naturalist.

There is a moral here and a lesson. But they're so obvious I don't have to spell them out. Not only had I lost my job, but I had simultaneously realized that my impossible dream was indeed impossible. Where would I get the tens of thousands of dollars it would cost to lead an expedition to Southeast Asia? Like Willy and Marvin, I had paid a terrible price for an ounce of wisdom.

I learned the worst way that wanting to be something doesn't make it so. Rootless and broke again, I returned to Columbus and signed back up at college. The G.I. Bill was reinstated, and I continued my studies. But another unexpected thing happened. I discovered that I knew much less than I thought about zoology and could pull only a C in the classes. Anatomy and classification bored me.

Thoroughly humbled, I found that the many books I'd read about Nature had prepared me not for a career as a naturalist, but as a lover and writer of literature. It was the mystery of life, and not life itself, which fascinated me. When I finally earned my master's, it was in literature and not zoology.

Why literature? Because when I had looked into the King Cobra's eyes those many times, my enchantment had not been generated by the biology of the beast, but by the essential mystery which made him so phenomenal and enigmatic in a world which seemed real and unreal at the same time. How could it be that we were both living beings made of the same substance and animated by the same soul, yet absolutely alone in our separate skins?

Staring into that cobra's eyes had also mystified me because I had to realize that the past is a distorted illusion persisting only in memory . . . and the future is just an ephemeral dream. Only the moment is real and never what we predicted or wished it would be. We are all condemned to exist in this eternal moment, and we are dragged with it – like a leaf in a river – as it moves further from yesterday and further toward an unknown shore of which we have no conception.

So what was this inexplicable bond between man and animal? What power or what chance lay underneath the persistence of life, struggle, love, birth, and death? The answer was not to be found in counting ribs or unraveling DNA strands. It was not to be found in dissections nor careful notes of feeding habits. It was only to be found in human intuition and awareness. Yes, the moment is dragging us toward a common destination. Yes, we are – man, beast, and plant – creatures made of light which mysteriously and ironically feed upon each other. But toward what apex, what unknown but wondrous height, is this eternal struggle between species heading?

As the years have passed, what seemed a tragedy when I was 23 has softened to a philosophical acceptance that the Curator was right to fire me. I am not scientific material, I confess. I am interested in how many vertebrae are in a serpent's spine, and I am equally interested in how often a crocodile must eat to swim one mile. But I am most interested in the phenomenon, the wonder, of the existence of life and how we beings can be so inextricably interconnected and obviously one growing organism, while at the same time so terrifyingly alone in our individuality.

Yes, the Curator had been right; I was not made for the scientific study of life. There is a deeper wisdom which I seek. That is why today I wander alone through the woods and wade down the open rivers, trying to absorb wisdom instead of equate formulas. Perhaps Nature has a reason for hiding her secrets? Nevertheless, I must continually seek the reasons, the purpose, of her ascension. But I suspect, as I wind my way through the forests and deserts of her world, that only when I am assimilated into her total being will I finally understand the wisdom I saw in the King Cobra's eye.

A Brief Bio of the Author

Richard Lee Fulgham was born in Dachau, Germany, in 1947.

Shipped to America in 1953, he has worked as a rattlesnake keeper, a Navy photographer, a forest towerman, a police photographer, a high school teacher, a professor of literature, a newspaper reporter, a freelance magazine writer and most recently an author of traditional print and online books.

He received his Bachelor's Degree in English Literature from Georgia's Columbus State University in 1972 and his Master's Degree in American Literature from the University of Kentucky in 1985. From 1985 to 1989 he was a writing and literature instructor at Kentucky's St. Catharine College.

From 1989 to 2000 he worked as senior writer of "The Lebanon News" in Lebanon, Va., and authored a history of southern Appalachia entitled APPALACHIAN GENESIS: THE CLINCH RIVER VALLEY FROM PREHISTORIC TIMES TO THE END OF THE FRONTIER ERA (TN: Overmountain Press, 2000), ISBN 1570720886.

His novel, MAN'S LAUGHTER; ANATOMY OF A MANHUNT, appeared in 2004. Soon to be available are his nonfiction novels LION: NIETZSCHE CONTRA CHRIST (true story of disturbed professor seeking "nobility in an ignoble world" and THE HOGS OF COLD HARBOR (a massive Civil War saga based on an actual diary in the author's possession). He lives in Bel Air, Md, with his wife Janet.

Printed in the United States
96904LV00002B/348/A